Toxic Risks

Science, Regulation, and Perception

Toxic Risks

Science, Regulation, and Perception

Ronald E. Gots, M.D., Ph.D.
National Medical Advisory Service
Bethesda, Maryland

Library of Congress Cataloging-in-Publication Data

Gots, Ronald E.
Toxic risks : science, regulation, and perception / Ronald E. Gots.
p. cm.
Includes bibliographical references and index.
ISBN 0-87371-510-1
1. Environmental health. 2. Toxicology. 3. Health risk assessment. I. Title.
RA565.G68 1992
615.9′02--dc20 92-20571
CIP

Direct all inquiries to CRC Press, Inc., 2000 Corporate Blvd., N.W., Boca Raton, Florida 33431.

PRINTED IN THE UNITED STATES OF AMERICA
3 4 5 6 7 8 9 0

Printed on acid-free paper

FOREWORD

Intended Audience

Intense specialization has produced major advances in knowledge and expertise, but it has also created a certain myopia—an inability to see the forest for the trees or the broader picture. An industrial hygienist may be so intent upon measuring levels of chemicals that he has little idea what the measurements mean or how they will be used. An occupational medicine physician, attuned to his patient's concerns, may blame the workplace with little scientific basis. Environmental toxicologists and regulators may be so involved with the number crunching of quantitative risk assessment that they may forget the underlying assumptions that provide uncertainties in their results and separate science from public policy. Regulators and legislators pressed by the public's demands for safe living environments and workplaces, assaulted by often conflicting input from experts, may have difficulty reaching reasoned decisions. Attorneys who must either enforce or challenge regulations or legislation need to understand the fundamental, as well as the superficial, basis for those rules and laws if they are to challenge or support them effectively. Risk communicators, a new specialty of the 1990s, have a special need to understand the biological and policy principles which underly risks and their perceptions. The general public, too, dissatisfied by the sensationalism or obviously-biased positions available in consumer-oriented publications, is interested in easy-to-read professional literature.

I was asked by a number of colleagues about bringing contributors in and editing rather than writing this book myself. While I greatly appreciated both their interest and their desire to preserve my family life, I decided not to do that for several reasons. First, I wanted to avoid the intense expert/specialist input which characterizes many edited books. Secondly, the very breadth of this message requires a sort of continuity or overview which is more amenable to single authorship. Finally, having been personally involved in so many of these areas for so long, there are things that I desire personally to convey to those who are interested.

The very intent of this multidisciplinary work is to simplify complex issues each of which has its own literature, texts and courses of study. Experts in each of these covered areas may see the treatment of their fields as over-simplifications or superficial. My hope is that for each chapter which is perceived by a reader as oversimplified, another will be illuminating. And, that readers from another field will have the same experience with a different set of chapters. That is the cross-disciplinary relevance which I am attempting to achieve. I will devoutly attempt to avoid inaccuracies or misrepresentations, but simplification is my goal.

Most professional-level books are written with a specific and narrow audience in mind. This one is not. It is an attempt to link a large range of disciplines. All of these disciplines (in which I include the concerned consumer) deal with these environmental and occupational health issues. Rarely do the individuals involved come together in a common understanding or have the time or the help to permit them to take stock of the forest before they become immersed in the trees. This book is dedicated to that overview--the grist of interdisciplinary communication. Some of the groups that need to understand each other better and for whom this book is written and dedicated include the following:

Attorneys
Corporate Health and Safety Officers
Environmentalists
Environmental Scientists and Industrial Hygienists
Journalists
Legislators
Municipal and State health and information officials
Physicians
Public Health Professionals
Public Policy Experts
Regulators
Risk Communicators
The Public-at-large

ACKNOWLEDGMENTS

I would like to thank my colleagues and family members (some of whom are also colleagues) for their important contributions to this effort. My thanks to Jason Gots and Dr. Barbara Gots for editorial assistance. Thanks to Dr. Joseph Gots, Karen Shaines, and Dee Karambelas for their writing and research contributions. Thanks to Tamar Hamosh for her proofreading and organizational reviews. And thanks to Ruth Ross-Hunley for her able document production.

CONTENTS

Chapter 1: Toxic Fears: A Case Study 1

Chapter 2: Science and the Pursuit of Biological Truths 15

Chapter 3: The Scientific Method 27

Chapter 4: Principles of Toxicology 39

Chapter 5: Selected Applications of Science to Environmental Toxicology 65

Chapter 6: Early Days of Environmental Concerns 87

Chapter 7: Cancer as the Catalyst: Scientists, the Media and Zealotry 91

Chapter 8: Environmental Chemicals: Common Perceptions Versus Scientific Knowledge 107

Chapter 9: The Rise of Toxin-Related Regulation 121

Chapter 10: Quantitative Risk Assessment: An Attempt to Link Science to Cancer Policy 143

Chapter 11: Litigation 153

Chapter 12: Dioxins and Agent Orange 171

Chapter 13: "Sick Buildings" 187

Chapter 14: Asbestos 209

Chapter 15: Lead 223

Chapter 16: Agricultural Chemicals and Cancer: The Case of Alar 247

Chapter 17: The Balance Between Risk and Reason 257

1
TOXIC FEARS: A CASE STUDY

INTRODUCTION

On April 3, 1987, the firefighters of Shreveport, Louisiana, responded to a call from the Louisiana State University Medical Center. By the end of that day, nearly 100 firefighters and police officers were involved in an event that would profoundly affect many of their lives. At the time of this writing, the aftermath of this episode is still in progress with perceived illnesses, paranoias, and lawyers doing battle on both sides and the city divided in support of and sympathy for the warring factions.

This story illustrates the central themes of this book. It underscores the popular misperceptions about chemical toxicity and the effects of those misperceptions. It exposes the depth of misinformation regarding basic toxicology, not only among the general public, but also in the medical community and even the regulatory agencies. It shows how medical quackery grows to fill the void of unresolved health fears. Finally, as a consultant to the city as this issue evolved, I gained intense personal knowledge of the facts and the data. With these, I was able to separate health effects and symptoms due to chemical toxicity from those due to emotional responses to a threatening event. Separating symptoms due to chemical concerns from those due directly to toxicity is essential to both assessment and treatment.

The story of the Shreveport firefighters is not intended to cast aspersions upon its participants, nor is it intended to be a journalistic report. Therefore, the events and players will be described purely as a means of illustrating the degree to which fears and perceptions rule behavior and how those fears and perceptions depart from actual data and scientific knowledge. Because that is the purpose, individual participants will not be identified by name nor will the facts be reported in a moment-by-moment reportorial fashion.

THE EVENT

On April 3, 1987, at 2 A.M., a call came in to the fire department from the Louisiana State University School of Medicine. When the firefighters arrived, they saw black smoke pouring from an upstairs room. Those who went up first were, for the most part, wearing protective equipment, including respirators, though some put their equipment on later. They arrived in the transformer room of the facility, whereupon an explosion occurred. It was not massive and caused little injury, but it was undoubtedly frightening. It was not accompanied by a fire, but by thick intense black smoke that poured from a smoldering transformer. The firefighters in the room noted a warning label on the transformer: "Danger, PCBs." Recognizing that PCBs (polychlorinated biphenyls) are considered hazardous, the firefighters were careful, at that point, to be fully protected. Shortly after the fire department appeared on the scene, the police department entered the picture, cordoning off the area, protecting the public from potential harm. Decontamination systems were set up to clean and detoxify affected clothing and equipment. In short, shortly after this event began, it was transformed from a typical fireman's duty—fighting fire—to managing a toxic disaster.

The Development of Health Concerns

Precisely how health concerns arose is not clear at this point, nor is it relevant to the purpose of this discussion. The essence of it is quite clear.

The "Danger PCBs" sign alarmed certain firefighters, some more than others. We know that emotional distress and fear affect people quite differently. That was clearly true in the case of the firefighters, some of whom reacted with particular alarm. In any situation like this one, there are profound emotional as well as physical responses at work. Involved individuals are understandably concerned about potential health effects. They want answers, they want help, and they will accept both from any perceived experts willing to give it to them.

The failure of all in authority to provide that help or acceptable answers ultimately transformed a toxicologically harmless episode into a major catastrophe. First, fears were heightened by the EPA's (Environmental Protection Agency) response. They cordoned off the area and, in Level I protective gear, decontaminated the building and its surroundings. Anyone seeing that exercise who had been on the scene during the height of the contamination would naturally have been frightened. Moreover, the EPA—those seemingly in the know—were not helpful in allaying concerns. They could not and would not reassure the firefighters and police officers. They simply gave them the regulatory answer about the hazards of PCBs. The EPA cautioned that "PCBs are highly potent carcinogens posing a clear and present threat to people's health." They had no answers for questions such as "Will I get cancer?" "Will I get sick?" "What do I do about this?"

By now, emotions were running high. The fire department contacted the local physician who generally handled routine health matters for it. As it happens, most family physicians have minimal if any training, experience, or knowledge in toxicology, particularly in toxicological matters relating to industrial chemicals. Medical schools do not offer these courses.[1] Consequently, he was not very helpful since he knew little about chemicals generally and even less about PCBs in particular. The physician held lay opinions about the dangers of PCBs, many of which were related to the intensity of regulatory activity concerning this chemical. Consequently, he was unable to provide accurate advice and answers to very understandable requests for information: "Will this hurt me?" "How do I know if I've been injured?" "What are the symptoms of poisoning?" "Will this exposure cause cancer?" "How can I be treated?" and others. Not knowing the answers, but sharing concerns about PCBs, the physician merely related what he knew. "PCBs are toxic and hazardous. They cause cancer." In that sense, he heightened concerns; after all, he was the "expert."

Understandably, some of the firefighters were now terrified, certain that they were going to get cancer, and that they had been poisoned. The degree of concern varied between individuals depending upon their personal psychological makeup and their perceptions about the degree of personal risk. Suffice it to say, however, there was widespread concern, bordering for some on panic. Many of them now began experiencing a variety of symptoms. Some had headaches, some chest pain, some skin complaints. Many complained of difficulty sleeping, nervousness, and trouble concentrating.

They began searching more frantically for physicians who could help them, who could treat their perceived disorders.

The firefighters were referred to a group in Dallas who purport to be experts in environmental poisonings and claim to be able to diagnose and treat people with such problems. They belong to a fringe medical group known as clinical ecologists, who will be discussed further in Chapter 2. These physicians measured PCB levels in the firefighters' blood and produced reports that included such statements as, "Your PCB levels were the second highest among all of the firefighters," and "PCBs are not natural chemicals. No levels are normal." While all of these statements were technically true, they were also misleading and frightening, geared more toward generating patients than toward relieving those who were worried unnecessarily. The fact is that PCBs are routinely found in everyone's blood (at very small levels) and the firefighters' levels were no higher than those of Americans in general. Therefore, these measurements should actually have led to reassurances because they demonstrated that the firefighters had not been "poisoned" at all in this incident.

At about this time, the firefighters read in one of their journals a discussion of PCB poisoning in firefighters and of a treatment center in California that claimed that it could eliminate PCBs from the system. Its technique, unproven, not medically accepted—another fringe medical procedure—involved aerobic exercises, a polyunsaturated fat diet, and high doses of vitamins. The level of fear and anxiety in many of these firefighters had, by now, reached a fevered pitch. Anxious to help, the city administrators agreed to pay for this very expensive California program.

THE FACTS OF THE EXPOSURES AND HEALTH COMPLAINTS

The City Becomes Suspicious

After approximately 15 of the firefighters had made the trip to California, the city began to question the value of the treatment for several different reasons. First, many of the firefighters continued to have symptoms after they had been treated. Second, the reports coming from the California group were not reassuring. They warned of future risk of cancer and other diseases. Third, the "treatment" was quite expensive, which heightened concerns about cost vs. benefit. Finally, some inquiries to the established medical and scientific community raised concerns about the legitimacy of this group and its methods.

Actual Findings

When I was asked by the city to investigate these "poisonings," I carried out a multiphased approach. First, I analyzed the world's scientific literature regarding PCBs: the effects of these chemicals on animals and human beings as defined by the numerous laboratory and epidemiological studies that have been performed. Important to the success of this phase was determining the quality of each study. Some studies are meaningful; some are useless and the resulting conclusions of little or no value.

Second, I reviewed the clinical data to determine what was wrong, if anything, with these firefighters and police officers; what their PCB levels were and whether they reflected poisoning, and what, if anything, the relationships were between the PCB levels and symptoms and the degree of exposure, as assessed by activities and locations during the event. In other words, were those who had the most symptoms the most exposed and were these the ones whose PCB levels were highest?

Third, I evaluated what these workers were told and what, as a result, they perceived to be true. If I was to meet with them, I had to understand both the scientific issues and the misconceptions I would have to allay.

Fourth, I reviewed the regulatory history of PCBs. Because much of the concern about PCBs has emanated from its regulatory history, that history and its relationship to scientific fact is an essential part of the perceived hazard.

Finally, I decided upon the best approach for communicating my findings to the firefighters and police officers. This was ultimately done in an open meeting which the city's television news media also attended. I shall abbreviate the findings of this investigation, but highlight certain elements that are critical to developing an understanding of the science, perceptions, and regulations of toxins.

SCIENTIFIC LITERATURE AND PCBs

For these "victims," there were substantial and relevant scientific data to allay their concerns. Numerous epidemiological studies evaluating the effects of PCBs had been performed, some in firefighters just like themselves; others in manufacturing workers making PCB—containing electrical equipment; and some in individuals heavily exposed in two large mass poisonings.[2-38] In electrical workers and in other firefighters, there were no adverse health effects. In fact, in the Shreveport firefighters, the blood and fat PCB levels were quite low (well within national norms), suggesting that little if any actually entered their systems as a result of the fire.[39] This information could have helped to reassure these firefighters, but it had been withheld from them.

Large-scale poisonings in Japan and Taiwan took place when rice oil became contaminated with PCBs. In these instances, the victims ingested enormous quantities of these chemicals over a protracted period of time. Their dosages were many orders of magnitude higher than the highest imaginable amount taken into the bodies of these firefighters. The blood levels of the victims, too, were high, sometimes as much as 200 times higher than the levels in the firefighters. The victims of this serious poisoning did develop some toxicological manifestations including certain skin problems, ocular disorders, and neurological and immunological effects. Even so, the effects were not, for the most part, terribly serious, nor were they life threatening. Twenty years later, the effects have still not been serious. There is, for example, no evidence of cancer or any other major disease. The scientific literature most relevant to these Shreveport firefighters was, therefore, quite reassuring. Acute poisoning was unlikely and long-term risk was essentially nonexistent.

Regulatory Response and Its History

Why, then, did the EPA alarm the firefighters so? Why weren't they reassuring? Why did they don such forbidding protective gear and engage in such extensive cleanup?

PCBs have followed the legislative and regulatory course begun by dichloro-diphenyl-trichloro-ethane (DDT) and later continued by Aldrin, chlordane, and the dioxins. All of these groups of chemicals share certain structural characteristics including multiple chlorinated aromatic rings. They are also characterized in common by protracted environmental fates and high lipid (fat) solubility. Because they do stay around for a long time and because they are potent animal carcinogens (at least in the case of dioxins, but somewhat less potent in the case of PCBs), they have been targeted as leading environmental hazards and central players on the regulatory hit lists.

In later chapters, we will probe the science versus perception issue with regard to the dioxins. The fact is that PCBs, like dioxins, have been labeled major environmental threats, elevated to this position as much by misperception as by scientific fact. Once a chemical is accorded this status, the agency's perspectives become molded by established perceptions until the realities and beliefs merge inextricably. A recent editorial in *Science Magazine*, "Excessive Fear of PCBs," highlighted aspects of these perceptions, pointing out, for example, "Many industrial workers were exposed to substantial amounts of PCB's...some respired a total of 15 grams or more. But the industrial exposure led to no known cases of cancer. Nevertheless, as much as $100 billion could ultimately be spent trying to remove the PCBs from the environment."[2]

The agencies treat PCBs as though they are one of the most dangerous substances known to man, even though the human scientific studies reveal that is obviously not true. Governmental agencies cannot be reassuring for a couple of reasons. First, the state EPA officials with whom I met had little scientific knowledge about the available PCB data. Therefore, for them, the regulatory history had become a surrogate for toxicological fact.

Second, the EPA is not in the business of providing reassurances. In fact, it is not in the business of dealing with individuals at all. It is only in the business of preventing potential risks in populations, a mandate that is, as in the case of the Shreveport firefighters, incompatible with the needs of the individual and his or her personal risks and, especially, his or her fears.

The Clinical Data

A review of the actual medical and toxicological records made it quite clear that not a single worker had been "poisoned," but many were suffering from fears and anxieties brought on by having been told that they were. Our first approach was to record all the blood and fat levels and to lay out in tabular form a variety of data. We catalogued the symptomatology compared with PCB blood levels. We charted symptoms as a function of where the individuals were during the event and correlated them with likely maximum exposure levels. Finally, we compared blood and fat levels with location during the event and likely exposure.

It was immediately apparent that none of the firefighters' PCB blood levels or fat levels exceeded those expected generally in the population. Second, they had none of the classic signs or symptoms of PCB toxicity; rather, the majority of reported symptoms were those associated with stress or anxiety. Third, there was no relationship between the level of PCBs in the blood or fat and the location of the individual during the event. Some of those who were in closest proximity to the transformer oil had levels higher, others lower, than those farthest away. These data supported the hypothesis that these were background levels and that no absorption of toxins had occurred at the site. Fourth, there was no relationship between blood and fat levels and the extent of symptoms. Again, this suggested that symptoms were unrelated to PCBs and were much more likely related to the extent of fear or anxiety in the individual. There was a correlation between the proximity of individuals to the center of the event and reported symptomatology. Clearly, the closer to the center of perceived contamination, the greater the perceived exposure. This had undoubtedly created a perception of intoxication with its expected emotional consequences in certain individuals. Graphs of severity of reported exposure versus PCB fat levels are shown in Figure 1.1.

This linear regression analysis compares PCB fat levels with exposure severity (based upon proximity). Sixty individuals were plotted. There was no relationship between exposure severity and PCB fat levels. In Figure 1.2 reported symptoms (graded by severity) are compared with PCB fat levels. Again, there is no correlation between the two.

PUTTING IT TOGETHER

The clinical information including the PCB levels in the blood and fat combined with the scientific toxicology and epidemiological literature made it abundantly clear that no poisonings had occurred. These individuals should have been told that immediately. Instead, they were led down a path of chemical hysteria charted by unscrupulous medical groups who profit by creating and exploiting such paranoias. The regulatory agencies, unable to provide reassurance to individuals, contributed in their own way to the hysteria. "After all," thought the firefighters, "if the government agency in charge of such things can't tell us everything is okay, then there really must be a problem." Their local physician provided no comfort or solace and was as convinced as his patients were that an unorthodox medical group was actually the expert.

Ultimately, these city workers became victims of a system that has encouraged fear of environmental hazards and toxins, but has not helped to put those fears into proper context. "When are these chemicals harmful?" "How are they harmful?" "Will they harm me given this exposure scenario?" These questions remained unanswered despite available, legitimate, fact-based information. During the next decade more such questions will arise. That an agent is regulated at the 10^{-5} risk level is of little relevance to an individual; nor is it terribly relevant as a matter of science. Until the lines between scientific knowledge, popular perceptions, and regulatory action are clearly elucidated, confusion will continue and more individuals will be victimized by unnecessary fears and bogus therapy.

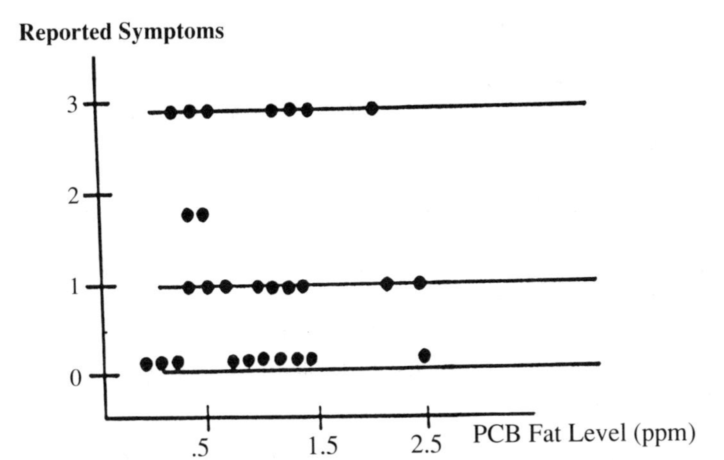

Figure 1.2 Linear Regression Analysis Comparing PCB Fat Levels with Severity of Symptoms

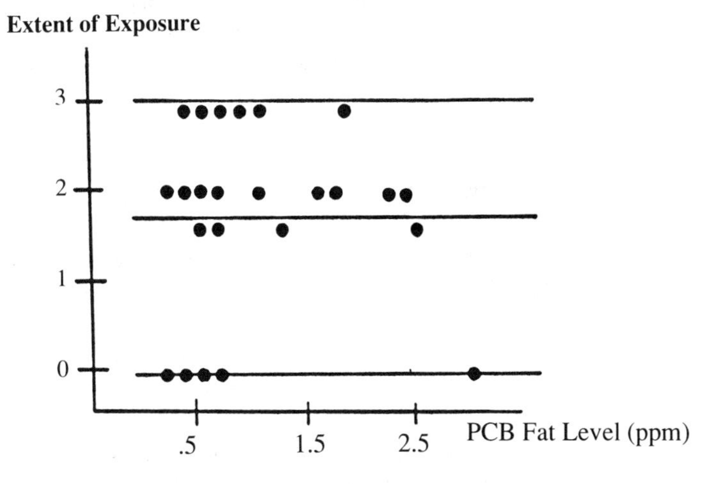

Figure 1.1 Linear Regression Analysis Comparing PCB Fat Levels with Severity of Exposure

REFERENCES

1. Institute of Medicine, *Role of the Primary Care Physician in Occupational and Environmental Medicine*, National Academy Press, Washington, D.C., 1988.

2. Abelson, P.H., Excessive fear of PCB's, *Science*, 253, 361, 1991.

3. Acquavella, J.F., Hanis, N.M., Nicolich, M.J. and Phillips, S.C., Assessment of clinical, metabolic, dietary, and occupational correlations with serum polychlorinated biphenyl levels among employees at an electrical capacitor manufacturing plant, *J. Occup. Med.*, 28, 1177, 1986.

4. Chen, P.H., Luo, M.L., Wong, C.K. and Chen, C.J., Polychlorinated biphenyls, dibenzofurans, and quarterphenyls in the toxic rice-bran oil and PCBs in the blood of patients with PCB poisoning in Taiwan, *Am. J. Ind. Med.*, 5, 133, 1984.

5. Chen, P.H., Wong, C.K., Rappe, C. and Nygren, M., Polychlorinated biphenyls, dibenzofurans and quaterphenyls in toxic rice-bran oil and in the blood and tissues of patients with PCB poisoning (Yu-Cheng) in Taiwan, *Environ. Health Perspect.*, 59, 59, 1985.

6. Chen, R.C., Tang, S.Y., Miyata, H., Kashimoto, T., Chang, Y.C., Chang, K.J. and Tung, T.C., Polychlorinated biphenyl poisoning: correlation of sensory and motor nerve conduction, neurologic symptoms, and blood levels of polychlorinated biphenyls, quaterphenyls, and dibenzofurans, *Environ. Res.*, 37, 340, 1985.

7. Chia, L.G. and Chu, F.L., Neurological studies on polychlorinated biphenyl (PCB)-poisoned patients, *Am. J. Ind. Med.*, 5, 117, 1984.

8. Eschenroeder, A.Q., Doyle, C.P. and Faeder, E.J., Health risks of PCB spills from electrical equipment, *Risk Analysis*, 6, 213, 1986.

9. Fischbein, A., Liver function tests in workers with occupational exposure to polychlorinated biphenyls (PCBs): comparison with Yusho and Yu-Cheng, *Environ. Health Perspect.*, 60, 145, 1985.

10. Fischbein, A., Rizzo, J.N., Solomon, S.J. and Wolff, M.S., Oculodermatological findings in workers with occupational exposure to polychlorinated biphenyls (PCBs), *Br. J. Ind. Med.*, 42, 426, 1985.

11. Fitzgerald, E.F., Standfast, S.J., Youngblood, L.G., Melius, J.M. and Janerich, D.T., Assessing the health effects of potential exposure to PCBs, dioxins, and furans from electrical transformer fires: the Binghamton State Office Building medical surveillance program, *Arch. Environ. Health*, 41, 368, 1986.

12. Hara, I., Health status and PCBs in blood of workers exposed to PCBs and of their children, *Environ. Health Perspect.*, 59, 85, 1985.

13. Hsu, S.T., Ma, C.I., Hsu, S.K., Wu, S.S., Hsu, N.H. and Yeh, C.C., Discovery and epidemiology of PCB poisoning in Taiwan, *Am. J. Ind. Med.*, 5, 71, 1984.

14. Hsu, S.T., Ma, C.I., Hsu, S.K., Wu, S.S., Hsu, N.H., Yeh, C.C. and Wu, S.B., Discovery and epidemiology of PCB poisoning in Taiwan: a four-year follow-up, *Environ. Health Perspect.*, 59, 5, 1985.

15. Huff, J.E., Moore, J.A., Saracci, R. and Tomatis, L., Long-term hazards of polychlorinated dibenzodioxins and polychlorinated dibenzofurans, *Environ. Health Perspect.*, 36, 221, 1980.

16. Kashimoto, T., Miyata, H., Fukushima, S., Kunita, N., Ohi, G. and Tung, T.C., PCBs, PCQs and PCDFs in blood of Yusho and Yu-Cheng patients, *Environ. Health Perspect.*, 59, 73, 1985.

17. Lawton, R.W., Ross, M.R., Feingold, J. and Brown, J.F., Effects of PCB exposure on biochemical and hematological findings in capacitor workers, *Environ. Health Perspect.*, 60, 165, 1985.

18. Lu, Y.C. and Wong, P.N., Dermatological, medical, and laboratory findings of patients in Taiwan and their treatments, *Am. J. Ind. Med.*, 5, 81, 1984.

19. Lu, Y.C., and Wong, P.N., PCB poisoning in Japan and Taiwan. Dermatological, medical, and laboratory findings of patients in Taiwan and their treatments, *Prog. Clin. Biol. Res.*, 137, 81, 1984.

20. Lu, Y.C. and Wu, Y.C., Clinical findings and immunological abnormalities in Yu-Cheng patients, *Environ. Health Perspect.*, 59, 17, 1985.

21. Luotamo, M., Jarvisalo, J., Aitio, A., Elo, O. and Vuojolahti, P., Biological monitoring of workers exposed to polychlorinated biphenyl compounds in capacitor accidents, *IARC Sci. Publ.*, 59, 307, 1984.

22. Masuda, Y., Health status of Japanese and Taiwanese after exposure to contaminated rice oil, *Environ. Health Perspect.*, 60, 321, 1985.

23. Ohgami, T., Nonaka, S., Watanabe, M., Yamashita, K., Murayama, F., Yoshida, H. and Masuda, N., PCB and PCQ concentrations in subcutaneous tissue from patients with PCB poisoning (Yusho), *J. Dermatol.*, 14, 25, 1987.

24. O'Keefe, P.W., Silkworth, J.B., Gierthy, J.F., DeCaprio, A.P., Turner, J.N., Eadon, G., Hilker, D.R., Aldous, K.M., Kaminsky, L.S. and Collins, D.N., Chemical and biological investigations of a transformer accident at Binghamton, NY, *Environ. Health Perspect.*, 60, 201, 1985.

25. Rappe, C., Nygren, M., Maklund, S., Keller, L.O., Bergqvist, P.A. and Hansson, M., Assessment of human exposure to polychlorinated dibenzofurans and dioxins, *Environ. Health Perspect.*, 60, 303, 1985.

26. Reggiani, G. and Bruppacher, R., Symptoms, signs and findings in humans exposed to PCBs and their derivatives, *Environ. Health Perspect.*, 60, 225, 1985.

27. Sahl, J.D., Crocker, T.T., Gordon, R.J. and Faeder, E.J., Polychlorinated biphenyl concentrations in the blood plasma of a selected sample of non-occupationally exposed southern California working adults, *Sci. Total Environ.*, 46, 9, 1985.

28. Sahl, J.D., Crocker, T.T., Gordon, R.J. and Faeder, E.J., Polychlorinated biphenyls in the blood of personnel from an electric utility, *J. Occup. Med.*, 27, 639, 1985.

29. Schecter, A. and Tiernan, T., Occupational exposure to polychlorinated dioxins, polychlorinated furans, polychlorinated biphenyls, and biphenylenes after an electrical panel and transformer

accident in an office building in Binghamton, NY, *Environ. Health Perspect.*, 60, 305, 1985.

30. Stark, A.D., Costas, K., Chang, H.G. and Vallet, H.L., Health effects of low-level exposure to polychlorinated biphenyls, *Environ. Res.*, 41, 174, 1986.

31. Stehr, P.A., Welty, E. and Liddle, J., Epidemiologic assessment of populations exposed to PCBs, in *Proceedings of Hazardous Wastes and Environmental Emergencies Conference*, Hazardous Materials Control Research Institute, Cincinnati, 1985, 158.

32. Stehr-Green, P.A., Welty, E., Steele, G., and Steinberg, K., Evaluation of potential health effects associated with serum polychlorinated biphenyl levels, *Environ. Health Perspect.*, 70, 255, 1986.

33. Stehr-Green, P.A., Ross, D., Liddle, J., Welty, E. and Steele, G., A pilot study of serum polychlorinated biphenyl levels in persons at high risk of exposure in residential and occupational environments, *Arch. Environ. Health*, 41, 240, 1986.

34. Takamatsu, M., Oki, M., Maeda, K., Inoue, Y., Hirayama, H. and Yoshizuka, K., Surveys of workers occupationally exposed to PCBs and of yusho patients, *Environ. Health Perspect.*, 59, 91, 1985.

35. Takamatsu, M., Oki, M., Maeda, K., Inoue, Y., Hirayama, H. and Yoshizuka, K., PCBs in blood of workers exposed to PCBs and their health status, *Prog. Clin. Biol. Res.*, 137, 59, 1984.

36. Wolff, M.S., Occupational exposure to polychlorinated biphenyls (PCBs), *Environ. Health Perspect.*, 60, 133, 1985.

37. Wuu, K.D. and Wong, C.K., A chromosomal study on blood lymphocytes of patients poisoned by polychlorinated biphenyls, *Proc. Natl. Sci. Counc. Repub. China [B]*, 9, 67, 1985.

38. Yoshimura, T. and Hayabuchi, H., Relationship between the amount of rice oil ingested by patients with Yusho and their subjective symptoms, *Environ. Health Perspect.*, 59, 47, 1985.

39. U.S. Environmental Protection Agency, Office of Toxic Substances, *Broad Scan Analysis of the FY82 National Human Adipose Tissue*

Survey Specimens, (Volume 1), Executive Summary, National Technical Information Service, Springfield, VA, 1986.

2
SCIENCE AND THE PURSUIT OF BIOLOGICAL TRUTHS

Humanity's pursuit for control over its own health has always been a contrast between folk beliefs, from leeches to evil humors, and scientific discoveries, from penicillin to heart transplantation. Great discoveries in science and new experiments often disprove medical "certainties" overnight. While the progress of science highlights errors in past perceptions, unproven beliefs invariably exist, and they probably always will. Popular demands for immediate explanations and cures for disease circumvent the pursuit of truth and inevitably lead to wide acceptance of unproven concepts. Among the many factors that have advanced our medical knowledge, none has been more important than the development of controlled studies. By contrast, its counterpart, anecdotal observation, is more prevalent, but less likely to yield lasting truths. Much of the knowledge medicine has acquired through controlled experimentation has, over time, been sustained, while most of what was believed due to uncontrolled observation has passed into the annals of medical history.

A STUDY IN CONTRASTS: ANECDOTAL OBSERVATIONS VERSUS CONTROLLED STUDIES

Sir Francis Bacon, a 17th-century philosopher and scientist, may have been the first to discuss the need for controls and the importance of recognizing and eliminating biases in studies. Bacon illustrated the need for controls with an example of sailors at sea during a severe storm.

> To ward off a shipwreck, they will make vows hoping to encourage God's favor. If they survive, they will take their votive offerings to church to satisfy those vows.

But, Bacon asked,

> Did their survival prove the success of their vows? What about those who were drowned despite offering the same vows?[1]

Those who drowned despite their vows are what today we would call "controls."

John Lind, an 18th-century ship's surgeon, was actually one of the first to apply this scientific method to a pressing clinical problem. Sailors, but rarely officers, suffered from the disease scurvy, known today to be caused by vitamin C deficiency. Lind observed the fact that the sailors and officers seemed differentially susceptible to the illness and considered how the two groups differed. The differences he identified were crowded living and diet. To test dietary factors he set up a controlled study: he divided the crew into six groups of two each. All were given a uniform diet, but each pair was given a different supplement: cider, vinegar, sea water, citrus fruits, or a variety of drugs. Remarkably, the two given the citrus fruits improved dramatically. He was thereby able to reach a preliminary conclusion that something in the fruit prevented scurvy. He tested the hypothesis in others and each time found that he could prevent or cure scurvy with citrus fruits.[2]

It would be nearly 200 years before ascorbic acid or vitamin C would be recognized as the specific deficiency responsible for the disease. But Lind had discovered both a cause (citrus fruit deficiency) and a cure. It was the use of that controlled study that permitted this historical and accurate observation. Without proper controls, Lind would never have been able to determine which food corrected the scurvy. Lind could, instead, have asked each sailor what he thought made the scurvy better or worse. He would undoubtedly have obtained numerous and varied answers, any of which he could have accepted as correct. Testimonials and personal beliefs often contrast starkly with the results of controlled experimentation.

Around the turn of the 20th century, the medical mainstream embraced two beliefs about the causes of disease. The first was known as "visceroptosis": a notion that held that a sort of falling down of the internal organs was responsible for a variety of ills. The second was that this visceroptosis set the stage for "autointoxication": a self-poisoning produced by toxins absorbed from one's own intestinal tract. Autointoxication, it was believed, could produce a host of ills ranging from symptoms such as fatigue and insomnia to bona fide diseases such as rheumatoid arthritis and diabetes. These beliefs weren't simply the odd ruminations of a crackpot fringe. They were espoused and popularized by the greatest physicians and surgeons of the day.[3]

Sir William Arbuthnot Lane, a renowned British surgeon, published 75 papers over a quarter of a century in which he argued persuasively for this theory of self-poisoning. What's more, he led an international movement of physicians and surgeons who treated this root cause of disease with major, high-risk surgery including removal of the entire large intestine. Less dangerous medical treatments were also employed, including laxatives, purgatives, and enemas. There exists, even today, some popular belief that clean bowels are one secret to good health—a vestige of the autointoxicant school of thought.

A central characteristic of the autointoxicant theorists was their reliance upon empirical, anecdotal observations to corroborate their theories. A series of patients with vague complaints or chronic illnesses would undergo the favored treatment of the moment and all would achieve some measure of reported improvement. These "empirical" observations would convince the physician that he was on the right track—that is, his diagnosis was correct and his methods were working. Missing from this approach was the key essential of scientific methodology—the use of controls. Medicine had not fully learned the lessons taught by Sir Francis Bacon and John Lind.

From the days of Hippocrates in the 6th century to the early 17th century, physicians adhered to the unsupported belief that through the body's various conduits flowed six different fluids, or humors. Diseases were thought to represent a disbalance of these humors.[4,5] Lack of any systematic scientific investigation kept that concept alive for nearly 1100 years.

A common practice through the mid-19th century was to drain the diseased humor through bloodletting, either by opening a vein or by using leeches to extract blood.[4,6,7] Interestingly, a certain few conditions for which this practice was employed actually benefitted from it. Specifically, heart failure—a disorder improved, at least temporarily, by a decrease in blood volume—was one. For the most part, however, conditions were worsened, and many patients (including, it is believed, George Washington) were killed by the practice. Personal anecdotal tales of survivals (from people who would have survived anyway) perpetuated this health care approach.

Personal testimonials and anecdotal observations are the antitheses of controlled observations. They fail entirely to eliminate the effects of chance, the natural progression of disease, the power of the mind, bias, and many other factors that may influence the outcome of an observation. Yet they are persuasive to most lay persons and even to many physicians. They have long influenced assessments of causes of diseases and their treatment and they still do. We've all seen instant anecdotes—crutches flung aside as the faith healer cradles the believer's head. Cancer sufferers fly to Mexico to buy drugs like Krobiacin and Laetrile, seduced by testimonials of miracle cures. Humans will invariably choose folk medicine when they are desperate for a cure and conventional medicine provides no relief for their illnesses.

EXPERIMENTAL DESIGN: DIRECT EXPERIMENTATION

The concept of controls is arguably the most important element of the scientific method, but it is only *one* element. Sir Francis Bacon may have been the first to articulate all the essential components of the scientific method. The key elements, he noted were observation, precision, experimentation, reasoning, and caution in reaching conclusions.[8]

Bacon's concepts of experimental design were beautifully translated to studies of disease causation by Robert Koch. Robert Koch's (1843-1910) elegant and simple experimental methodology proved conclusively and for the first time that bacteria caused disease.[9] Prior to Koch's seminal studies, scientists could not determine what role bacteria played in infections. The prevailing notion was that they grew in diseased tissue—a result of an illness, rather than its cause. In order to test the hypothesis that bacteria actually caused the disease, Koch had to show that bacteria taken from one animal could reproduce the disease in another and that it could do so after every transplantation. Koch removed bacteria from an infected site, cultured the bacteria in a culture medium, and infected another animal, producing the same condition. He then recultured the organisms from the second animal and repeated the experiment. This simple but elegant study proved that the bacteria caused the infection. The resulting elements of the study became known as "Koch's Postulates." These are

1. The organism must be shown to be present in every case of the disease;

2. The organism can be cultured in pure culture;

3. Inoculating the animal with culture will reproduce the disease; and

4. The organism can be recovered from the inoculated animal and grown again in pure culture.

Due to the proper application of scientific methodology, the truth of Koch's observations is clear and enduring.

EXPERIMENTAL DESIGN: OBSERVATIONAL (EPIDEMIOLOGICAL) STUDIES

One of the most elegant of the early epidemiological studies was performed by John Snow, a British physician who linked the then—frequent cholera outbreaks in London to sewage-contaminated water.[10] Snow noted that outbreaks of cholera in London appeared to be concentrated in areas where two companies supplied water to the homes—the Lambeth Company and the Southwark and Vauxhall Company. He hypothesized that contaminated water from the Thames River was responsible; both companies drew their water from sites near sewage discharges.

At first, this was only a hypothesis, but then the opportunity came to test his belief using elegant epidemiological methodology. In 1854 there was a terrible outbreak of cholera that killed approximately 500 people in ten days. This occurred shortly after the Lambeth Company changed its water source to an area of the Thames River more distant from the site of waste discharge. Snow went from door to door to each house in which cholera had occurred and determined the source of each home's water supply. By gathering records he was readily able to determine the total number of homes supplied by each water company. His compilation of the cholera deaths according to water supply source is shown in Table 2.1.

Table 2.1 Cholera Death Rates in London, 1853-1854, According to Water Company Supplying Actual House

Water Company	Number of Houses	Cholera Deaths	Deaths per 100,000 Houses
Southwark & Vauxhall	40,046	1263	315
Lambeth	26,107	98	37
Rest of London	256,423	1422	59

Source: J. Snow, on *The Mode of Communication of Cholera* (2nd ed.). Churchill, London 1855. Reproduced in *Snow on Cholera*, Hofner, New York, 1965.

Snow discovered that the frequency and distribution of the disease was far greater in houses supplied by Southwark and Vauxhall Company, whose source was near the waste discharge. With this data, Snow made critical and novel epidemiological observations, identifying both frequency of disease and a likely causal factor.

OBSERVATIONAL BIASES AND ERRONEOUS BELIEFS AND PRACTICES

An operation known as the Vineberg procedure was used to relieve anginal pain until 1970, when coronary artery bypass procedures were first developed. The Vineberg procedure involved the dissection of an artery away from the inside of the chest wall following which it was placed in a tunnel made in the heart muscle itself. Surgeons who performed thousands of these procedures in the late 1960s swore by the operation. Innumerable testimonials "proved" that it relieved pain.

Operations, even today, come about by popular acclamation, often with little scientific investigation. The Vineberg procedure, however, was not studied in a scientifically controlled fashion until 1970. At that time, a group of surgeons, uncertain about the merits of this popular operation, performed a controlled study. They took a group of 100 patients with anginal pain and randomly divided them into two groups. One group got the Vineberg operation, the other a "sham" procedure, an operation in which the chest was opened and closed again, but the artery was not actually dissected away from the chest wall or stuck into the heart muscle. Such an experiment, I might note, could probably not be performed today because of modern ethical and liability concerns. Today, therefore, it is more difficult to employ controls to check the medical validity of an operation. The Vineberg study was conducted in a meticulously controlled and "blinded" fashion. The patients were randomly divided between two identical groups, differing only in the operation that each received. The patients were "blinded" (kept unaware of which operation had been performed in whom) as were the doctors grading the postoperative pain in both groups ("real" and "placebo").

The results were striking. Approximately 70% of both groups experienced dramatic relief of their postoperative pain. Critically, there was no difference between the two. When the article describing this study was published in 1970, the implementation of the Vineberg procedure came to an abrupt, overnight halt.

This story further illustrates the essentiality of controls, but it illustrates something more: one of the key reasons that controls are necessary. Patient bias is an important confounder of experimental results, but it is not always obvious or apparent. In this instance, one might have opined that controls were unnecessary because these patients had such persistent and severe chest pain that relief would "prove" the value of the procedure. That turned out not to be the case, because belief and the anticipation of relief can be extremely powerful healers.

MODERN ANTISCIENTIFIC SENTIMENT AND ANECDOTAL ILLNESS

Even today, the power of popular belief compels both the public and physicians to adopt odd notions about causes of and treatments for illnesses. The yeast theory of disease and current chemical allergy theories are two of the more striking examples.

The 1980s was the decade of syndromes: chronic fatigue syndrome, sick building syndrome, premenstrual syndrome, 20th-century syndrome, and multiple chemical sensitivity syndrome, among others. These labels were provided to explain a variety of physical and emotional complaints not explained well, or to the satisfaction of the sufferer, by available diagnostic methods. While under some circumstances some of these labels may explain accurately an individual's problem, for the most part diagnoses are difficult, if not impossible, to make, and some of these illnesses may not exist at all. In this category are chronic yeast infection syndrome, 20th-century syndrome, and multiple chemical sensitivities syndrome.

Chronic candidiasis (yeast infection) syndrome is a belief popularized in the books *The Yeast Connection* and *The Missing Diagnosis*.[11,12] This belief holds that all sorts of physical complaints and dysfunctions can arise in women as a result of the growth of yeast in the vagina. In the typical style of the crusader bucking the establishment, accepting the derision of their colleagues, each author uses anecdotes to validate his notion. Such a theory is attractive because many people simply don't feel well for reasons that are not invariably determinable. Moreover, of course, it is human nature to seek an explanation for not feeling well. I have a feeling that theories like the yeast concept are growing in popularity because they are hard to disprove. The medical establishment can rail against them, but if it hasn't disproven them, they stand as accepted in the public mind. A third reason that such a theory is attractive is that it contradicts mainstream medical thought and, particularly, scientific thought. The public is not positively disposed these days toward science and its practitioners. Perhaps we feel let down by failed promises—the war to cure cancer—and by the most visible (highlighted) products of technology—depleted ozone layers, acid rain, and polluted water and air.

It so happens that the theory of yeast infection as a cause of all sorts of ailments can readily be tested with less draconian measures than a "sham" operation. Because it had attracted a large following, it was tested. The candida organism, which causes yeast infections, can be killed with the drug Nystatin. In a controlled study designed to test the yeast theory, a group of 42 women who had been diagnosed as having this syndrome were studied in a controlled, blind fashion.[13] The women were given either a placebo (a "sugar pill,") or Nystatin. They were switched from a placebo to the active drug and vice versa, never knowing which they were taking.

The results were clear and compelling: the Nystatin reduced vaginal itching—a well-understood and expected symptom of candida vaginitis. It did not, however, differ from the control in reducing other symptoms such as gastrointestinal distress, respiratory symptoms, and depression-like complaints; all of which had been popularly linked to this "yeast syndrome." Both the placebo and the Nystatin reduced these symptoms by about 35%. This demonstrated both the psychological component of these symptoms and the placebo effect in their reduction, as well as the lack of involvement of the yeast infections in the process. Once again, controlled studies and competent experimental design had disproved a widely held belief.

More difficult to study is the growing perception that all sorts of chemical exposures from various environmental sources are producing all kinds of illnesses. Such conditions have been called by a variety of names "20th-century disease," "immune dysregulation," and others. Medical practitioners who espoused this theory were known as "clinical ecologists" and now call themselves "environmental physicians."[14-17] Their beliefs appeal to the same instincts as those beliefs that perpetuated the yeast theory, but with a feature that enhances this appeal. By blaming the environment, and particularly man-made chemicals, clinical ecologists couple popular antitechnological sentiments with an explanation for otherwise unexplained physical and emotional dysfunction.

In every way, this movement is the modern equivalent of the autointoxication theory, enjoying a growing number of professional adherents and lay believers. Like the autointoxicant theories, it explains otherwise disjointed symptomatology by invoking a common cause: toxins. Only the source of the toxins differs; a hostile external chemical environment has replaced poisons released by one's own large intestine. "Environmental illness" continually proves its own validity through anecdotal tales and testimonials. And, as were the autointoxicant theories, it is difficult to study or to disprove. Nonetheless, attempts have been made to study systematically some of today's "victims" of multiple chemical sensitivities.[18-22] None of the systematic studies has found either consistent objective physical similarities among those diagnosed or identifiable external causes, but each has identified similar psychological characteristics and makeup among patients. The best current data suggest that certain psychological disorders and qualities predispose individuals to develop symptoms and to seek out these environmental physicians to explain those symptoms.

There is one very interesting difference between current unproven beliefs about disease causation and those of prior centuries. Formerly, physicians were treated as omniscient sages. It was they who promulgated theories and made all medical decisions. The 1960s brought empowerment of the public, holistic medicine, and a decline generally in public esteem of science and medicine. With the holistic health movement came a redefinition of who controlled the public health. The public became actively participatory, with the help of sympathetic physicians, in defining health issues including causes of diseases. Thus, each individual became a personal laboratory, "understanding" his or her own body, "knowing" how various factors or substances influenced it. In effect, the science of medicine gave way to these newly encouraged personal testimonials and anecdotal reports, and physicians became more mere chroniclers of these popular anecdotes than students of disease causation. The net effect of this "enlightenment" on the science of medicine has been as negative as the "discovery" of autointoxication. Gross misperceptions existed then about the causes of disease and they continue to exist today. But the authorship of those perceptions is important to the central theme of this book, for when health perceptions spring forth from the public at large, they have the immediate force of political pressure behind them. Where chemicals are concerned, this is translated into legislation and regulation.

As antiscientific sentiments spread, personal misconceptions about human health spread also. In the Middle Ages evil humors were thought to cause most diseases. Today, environmental chemicals have replaced them in the public mind. Neither is true, but how do we know? The scientific method is the means by which we explore the truths of the biological sciences. The next chapter will delve into that method in further details.

REFERENCES

1. Bacon, F., The New Organon, in *The New Organon and Related Writings*, Anderson, F.H., Ed., The Liberal Arts Press, New York, 1960.

2. Lind, J., A treatise of the scurvy (Edinburgh, Sands, Murray and Cochran, 1753), reprinted in *Lind's "Treatise on Scurvy,"* Stewart, C.P. and Guthrie, D., Eds., The University Press, Edinburgh, 1953, 145-148.

3. Lambert, E.C., *Modern Medical Mistakes*, Indiana University Press, Bloomington, 1978.

4. McGrew, R., *Encyclopedia of Medical History*, McGraw Hill, New York, 1985.

5. Ogden, M.S., Guy de Chauliac's theory of the humors, *J. Hist. Med.*, 24, 272, 1969.

6. Bryan, L.S., Blood-letting in American medicine, 1830-1892, *Bull. Hist. Med.*, 38, 516, 1964.

7. Niebyl, P.H., The English blood-letting revolution: modern medicine before 1850, *Bull. Hist. Med.*, 51, 464, 1977.

8. King, L.S., *Medical Thinking: A Historical Perspective*, Princeton University Press, Princeton, 1982.

9. Koch, R. Ueber bakteriologische Forschung, *Verh. Int. Med. Cong. [Berlin]*, 35, 1892, 1980.

10. Snow, J., *On the Mode of Communication of Cholera*, 2nd edition, Churchill, London, 1849. Reproduced in *Snow on Cholera*, Hofner, New York, 1965.

11. Crook, W.G. *The Yeast Connection: A Medical Breakthrough*, 3rd edition, Professional Books, Jackson, TN, 1989.

12. Truss, C.O., *The Missing Diagnosis*, 2nd edition, C.O. Truss, Birmingham, 1986.

13. Dismukes, W.E., Wade, J.S., Lee, J.Y., Dockery, B.K. and Hain, J.D., A randomized, double-blind trial of nystatin therapy for the candidiasis hypersensitivity syndrome, *N. Engl. J. Med.*, 323, 1717, 1990.

14. Dickey, L.L., *Clinical Ecology*, Charles Thomas, Springfield, VA, 1976.

15. Randolph, T.G., An ecologic orientation in medicine: comprehensive environmental control in diagnosis and therapy, *Ann. Allergy*, 42, 330, 1968.

16. Randolph, T.G., Human ecology and susceptibility to the chemical environment, *Prog. Allergy*, 19, 779, 1961.

17. Finn, R., and Battcock, T.M., A critical study of clinical ecology, *Practitioner*, 229, 883, 1985.

18. Black, D.W., Rathe, A. and Goldstein, R.B., Environmental illness: a controlled study of 26 subjects with 20th century disease, *JAMA*, 264, 3166, 1990.

19. Brodsky, C.M., Multiple chemical sensitivities and other environmental illness: a psychiatrist's view, in *Workers with Multiple Chemical Sensitivities*, Cullen, M.R., Ed., Hanley & Belfus, Inc., Philadelphia, 1987.

20. Terr, A.I., Multiple chemical sensitivities: immunological critique of clinical ecology theories and practice, *J. Occup. Med.*, 2, 683, 1987.

21. Terr, A.I., Environmental illness, a clinical review of fifty cases, *Ann. Intern. Med.*, 146, 145, 1968.

22. Scientific Task Force Report, Clinical ecology: a critical appraisal, *West. J. Med.*, 239, 1986.

3
THE SCIENTIFIC METHOD

In the preceding chapter, we discussed certain historical precedents in science. Examples from history show us science gone wrong. When it does, we reach odd conclusions about the causes of illnesses and treat people with varieties of nasty and dangerous methods. One thing is inevitable. Decisions about causes of diseases or their treatment that are derived from bad science never last. Well-designed investigation uncovers truths that, because they produce results time and time again, become part of the permanent fabric of knowledge.

How are such truths uncovered? In this chapter we will explore further the nature of such studies. If we are to separate scientific concepts based upon enduring principles from those resting on quicksand, we must understand the differences. Proper scientific methodology is not subject to the whim of anyone who would conduct a study. There are good studies and bad: the first providing meaningful information; the second providing information that is, at best, useless and, at worst, misleading. Unfortunately, poorly designed studies do get published. Thus, the scientific literature is an admixture of both well-conducted and poorly conducted research, each making an impact.

To distinguish science from perceptions (the intent of this book), we must have a basic understanding of the elements of science. A brief journey into the characteristics of scientific investigation follows.

AN EXPERIMENT

An experiment is designed to test an hypothesis. For example, if we want to know whether a drug causes an animal's heart rate to increase, we can find out by giving the drug to the animal and comparing the heart rates of treated animals with those treated with a control drug. If we want to confirm the effects in people, we give the drug to people. In such an experiment we directly set up the parameters of the study. Test subjects get the drug, but nothing else. Control subjects get a placebo.

This type of study is direct experimentation. It contrasts with observational studies such as the epidemiological studies that will be discussed more extensively in Chapter 5. Epidemiological studies follow the basic parameters of experimental design, but they differ from direct experiments in that we often have less control over the parameters. Many such studies examine events that have already occurred. Thus, the parameters are set up for us. We are forced to "observe" them. In such cases, we have far less control over the circumstances of the experiment. We might, for example, want to evaluate the role of diet in heart disease. We may be committed to evaluating individuals whose diets we did not control, but who already have heart disease. We can set up an observational study in which the group with heart disease is compared with a group without heart disease to correlate past dietary habits with the incidence of disease. In this instance the researcher has no direct control over the factors being studied, but merely over the study methodology. In the case of drug testing, the researcher sets up the parameters of the study. Sometimes the term "basic science" is used to reflect the extent to which the researcher has control over the parameters under investigation.

I remember my Ph.D. adviser telling me about bitter arguments between him, a scientist who worked with homogenized tissues and even whole organs, and his brother, a "pure" biochemist who worked in clean enzyme systems. Sam's brother insisted that his was "basic" science because parameters such as chemical compositions of test and control systems could be controlled precisely. Sam would reply that to a nuclear physicist who worked with subatomic particles, his brother's enzymes and other proteins would seem entirely too complex and far too messy to produce any meaningful information. "Basic," it would seem, is in the eye of the beholder.

Few would argue, however, that the less control there is over the experimental environment, the more assumptions will have to be made and the weaker the conclusions. If I attempt to correlate a specific factor with a child's behavior, I do so at great peril, because so many complicating factors are at work. If, by contrast, I test bacteria in a culture tube to see whether they will be killed by a particular antibiotic, I can readily control the study and arrive at a clear-cut and reproducible answer. In the study of behavior, we may find a relationship that may suggest a possible cause, but never prove one. In the antibiotic study, we can learn conclusively and absolutely whether the drug kills the organisms.

In the case of toxicology, both methodologies are applied: controlled experimentation, in which answers are reproducible and clear-cut, and observational studies in which answers are far shakier and much more uncertain. The direct toxicity of a chemical may be tested with a laboratory-based animal study. The investigator can determine directly and unequivocally how much it takes to kill the animal or how much it takes to cause liver failure. By contrast, workers may be evaluated in observational studies to assess relationships between chemicals and cancer or between chemicals and more subtle effects like cognitive dysfunction, intelligence, or immune function.

These studies are invariably fraught with significant methodological design problems. They rarely lead to clear-cut answers. Yet they are growing in popularity, because they are designed to investigate the most common questions of today. Specifically, they are the only way to address directly the toxicological questions posed by regulatory agencies.

INDEPENDENT AND DEPENDENT VARIABLES

The central elements of any experiment are the independent variable and the dependent variable. The first is the hypothesized cause. The latter is the hypothesized effect. Let us continue with the new drug as an example. Assume that we want to know whether the drug that we have just made in the laboratory causes the heart rate to speed up.

This is a simple experimental design in which a treatment is tested and an observation assessed. Figure 3.1 illustrates these variables. The question is, does the drug (X) cause the heart rate (Y) to speed up?

Central to any direct experimental design is this concept of cause and effect. We want to know whether what we are testing produces the effect for which we are looking. The cause and effect are the independent and dependent variables:

> Independent variable—a factor that is believed to influence the outcome we wish to study: the cause.
>
> Dependent variable—the variable whose change we are studying: the effect.

Thus, in our heart drug example, the drug is the independent variable; the heart rate is the dependent variable.

CONTROLS

The simplest way to do this study would also lead to an incorrect conclusion. We could check the animal's heart rate, then give the drug and then check the heart rate again. We might then conclude that anychange, either higher or lower, was due to the drug. What is wrong

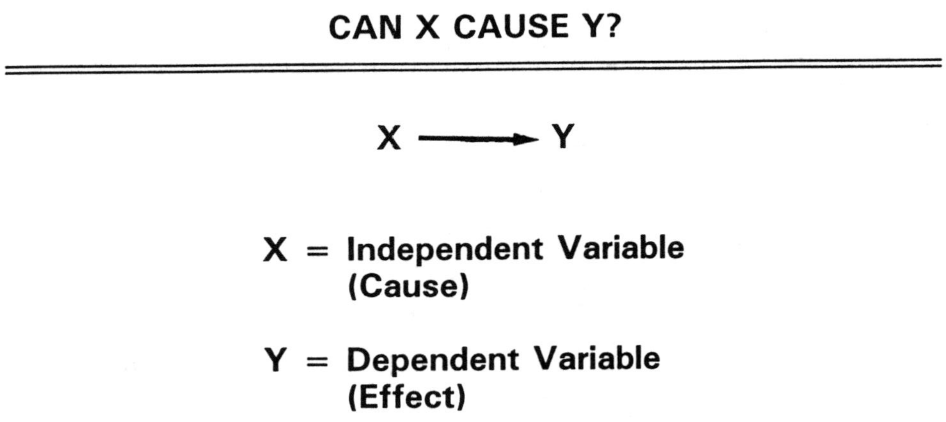

CAN X CAUSE Y?

X = Independent Variable
(Cause)

Y = Dependent Variable
(Effect)

Figure 3.1 Experimental Design

with this approach? What is wrong is that it does not enable us to distinguish between the effect of the drug itself and the effect of some other factor: handling the animal and making it nervous, for example. In order to separate that effect we need controls. We must handle a matched group of animals identically in every way except for the drug. They must be given an inactive agent (a salt solution, for example) known as a placebo.

Proper experimental design demands comparison, which means controls. What we are studying has to be compared with something. The essentiality of controls is highlighted by studies of pain relievers. One might think that if a medication produces pain relief in a patient following surgery—when the pain is obviously real and severe—such a patient's testimonial would be proof positive that the medication worked. As it turns out, if people suffering serious postoperative pain are injected with water, devoid of any drug, 30-40% will be relieved of that postoperative pain. Thus, more than one-third of the time, faith alone can relieve postoperative pain; hence the need for a comparative group or controls. Failure to employ controls has led to innumerable wrong conclusions about both causes of illnesses and effects of treatment. Examples are the autointoxication theories and the Vineberg operation, described in the previous chapter.

Certain laws require controlled scientific studies before claims of health effects can be made. The Food, Drug and Cosmetic Act demands carefully controlled studies before approval of drugs. A company cannot claim that a new antibiotic will cure streptococcal meningitis and market it with such a claim without having proven that to be so with controlled studies.

MINIMIZING BIASES AND CONFOUNDERS

Another critical component of proper experimental design is the minimization of bias. Controlling biases means reducing (it is hoped to zero, but that is rarely possible) the effect of factors other than the one being studied on the outcome. It is much easier to do this in direct studies than in observational studies in which all of the confounders may not even be known. One way of minimizing biases is through randomization. We randomize so that the control and the experimental group share as many of the same characteristics as possible. For example, we don't want to test the Vineberg versus the sham operation by performing the sham on individuals with less severe disease or with pain of more recent duration. If we did that, we wouldn't be sure at the end whether the results were due to the operation or to other characteristics of the two groups that made them respond differently. For example, even cultural factors may lead to a greater perception of pain relief in one group versus another. Randomization ensures that the cultural influences of the experimental and control groups are evenly distributed.

Blinding is also essential to the minimization of bias. We assign the patients to blind groups to control patient bias. Remember how effective the mind can be when it comes to pain. The researchers did not want patients to know that they had a "fake" operation. Their pain might not have been relieved simply because they didn't expect it to be. The researchers also did not want the patients who had the "real" operation to know that, because their pain relief might be due to their expectation of improvement. By maintaining uncertainty in all members of both groups, there could be no differential influence. Neither group knew which operation it had had.

Observer bias, too, needs to be controlled. Degrees of pain relief are qualitative judgments that can be influenced by both the patient and the observer. The observer, knowing which operation the patient had, could even influence (often subconsciously) the patient, cluing him in to the right answer, thereby affecting the results. This bias can be controlled by not letting the observer know the patient's group. If he or she doesn't know which operation the patient had, he or she cannot possibly influence the reporting about the extent of pain relief. The influence of biases is shown in Figure 3.2. Factors A, B, C, D, and so on are confounders and biases that influence the outcome (Y) under study. If these factors are not controlled, any outcome presumed to be due to the cause that is the subject of the study (X) may actually be due to one of these confounders. The minimization of biases is illustrated in Figure 3.3. Here, the confounders and biases are controlled because both the experimental group and the control group are influenced equally by them.

DOUBLE BLINDING

Double blinding is the term applied to a study in which neither the test subjects nor the observer knows whether an individual belongs to the test or the control group. All Food and Drug Administration (FDA) drug approvals require double-blind studies. If Prozac, an antidepressant medication is being tested, a control group is treated with coded, unmarked pills. Patients are questioned about relief of depression and

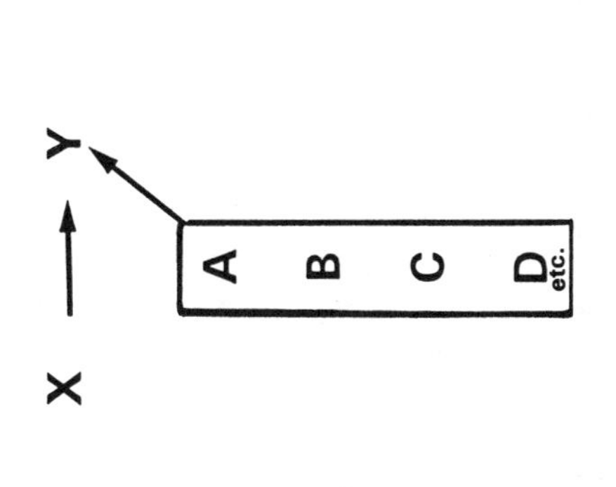

Figure 3.2 Confounders + Biases

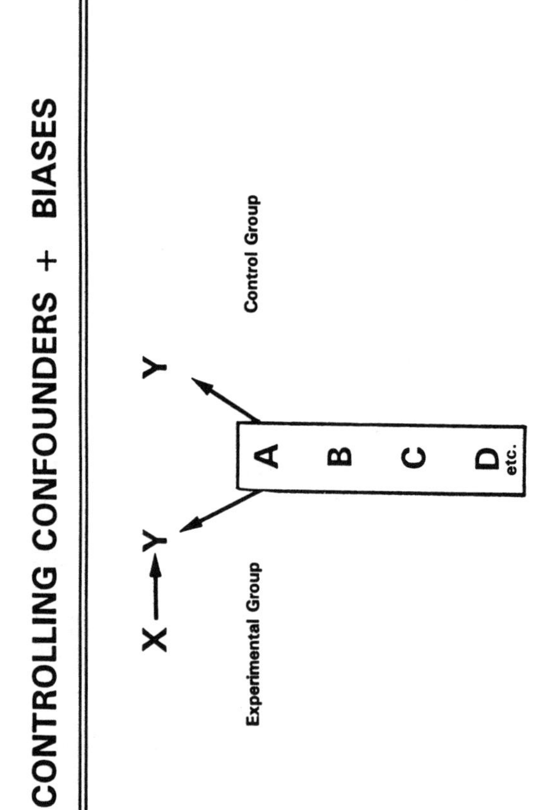

Figure 3.3 Controlling Confounders + Biases

symptoms or side effects. All of these are subjective responses, which increases the need for effective double blinding. Neither the recipient of the pills nor the observers recording the patients' responses know whether a given subject has taken a placebo or the Prozac. Only later, after the recordation is completed, are the codes broken and the medicated group compared with the control group.

Another example of a double blind study, described in Chapter 2, explored the cause of symptoms. To study the hypothesis that yeast infection produced a variety of symptoms, a test group was compared with a control group. The control group was treated with a placebo; the test group with the drug Nystatin, which eradicates yeast infections. Neither the patient nor the observer knew which was which when the participants recorded their symptoms. At the end of the study, the groups were compared and it was found that the treated group had no more change in symptoms than did the control group, thereby showing that the presence of yeast was not producing symptoms. This double-blind study was able to disprove that theory.

By contrast, the theory that yeast infection did produce widespread symptoms had developed because of anecdotal observation. Patients and doctors believed that symptoms were associated with the infections. They perceived a connection. Such perceptions are commonly proven to be in error once they are formally tested.

STATISTICAL ANALYSIS

Establishing the controls, blinding, and minimization of biases are done up front when the study is organized. These are elements of study design. Once the data are collected, they must be analyzed to see what they show. Analysis of data cannot be done simply by "eyeballing" them to reach conclusions. An example may help show why.

Assume that we are testing a medication to be used by Autoimmune deficiency syndrome (AIDS) sufferers to determine whether it can prevent infections from developing when it is taken prophylactically. Eight hundred patients are divided at random into two groups. One group is treated with a placebo, one with the drug. The experiment is double blinded. At the end of the test period, say one year, we find that 26% of the treated group and 18% of the control group were infection free. The question now is, Did the drug work? That, of course, depends in part upon how we define working. After all, a 26% infection-free rate may not be considered very effective. But assume that we are defining working in the pure scientific sense: Did it reduce infections more effectively than the placebo drug? The numbers show us that we are comparing 26% out of 400 with 18% out of 400. There is no way of eyeballing the numbers for us to know whether that is a meaningful difference or a chance difference. We therefore must apply statistical analyses to the data to answer that question. When we do, we find that the probability of finding differences of this size from random occurrence alone is less than one in one hundred (0.01). This suggests that this difference was not purely chance, but that the drug actually did reduce the frequency of infections.

Statistical analysis permits us to draw conclusions from data. All formal statistical reasoning is based upon laws of probability. These are laws used to explain the phenomenon of randomness or the likelihood that something will occur through chance alone. Thus, statistics help us to decide whether an observed phenomenon was the result of chance or whether it was more likely associated with a specific cause.

My purpose in this discussion is not to present statistical methodologies. For that I refer the reader to a number of excellent references at the end of this chapter. The specific approaches to statistical analysis can be quite complex. In fact, statisticians regularly debate over the applications and even the value of particular statistical methods. This is especially so for the ones commonly used in complex inductive analysis, that is, in epidemiological studies. The aim in this brief discussion is simply to round the edges of experimental design—to illustrate the fact that proper conclusions cannot be reached without formal data analysis and that statistics is the language of such analyses.

Statistical analyses tell us whether relationships, that is, between the independent and dependent variable, are significant or not. Since significance or lack of significance is often counterintuitive, statistical tests may contradict our impressions. An example is the concept of the "law of averages." It is commonly believed by gamblers (but not professional gamblers if they are to continue in business) that runs of bad luck are necessarily followed by a string of good luck. Five tails coin tosses in a row, must be followed by a heads according to the law of averages. The fact is that each coin toss is independent of the one before it. Assuming that the coin is not a weighted or trick coin, each toss has a 50:50 chance of being heads or tails regardless of what came before. There is no such thing as the "law of averages" controlling the outcome of the next coin toss. This common belief in "law of averages" and failure to understand rudimentary probabilities leads to common misperceptions in environmental toxicology. When a number of workers or neighbors develop what is perceived to be a similar and rare disease, human nature searches for a common explanation. People believe that the "law of averages" mitigates against such illnesses occurring by chance. In fact, these clusters, as they are known in epidemiology, are statistically explainable phenomena, occurring randomly and by chance. Only through systematic, scientific analyses of such occurrences, with proper controls and careful statistical study, can one know whether the intuitive perceptions about relevance have any demonstrable validity. Very commonly, it is this intuition, unsupported by study, that drives public calls for action and, ultimately, public policy.

SUMMARY

Scientific investigation is not a haphazard process. In order to discover biological truths, observations must take place in a systematic way—one that is known to produce reproducible and meaningful answers. Experimental designs must consider carefully ways to maximize the observation of true relationships while the influence of irrelevant factors is minimized. Data must be assessed through proper application of correctly-selected statistical approaches.

Perceived significance alone is insufficient and is, in fact, often wrong, since intuition is neither systematic, rigorous, nor mathematical for most individuals.

GENERAL REFERENCES

Hennekeus, G.H., and Buring, J.E., *Epidemiology in Medicine*, Little Brown & Company, Boston, 1987.

Moore, D.S., *Statistics: Concepts and Controversies*, W.H. Freeman and Co., New York, 1985.

Paulos, J.A., *Innumeracy: Mathematical Illiteracy and Its Consequences*, Hill and Wang, New York, 1988.

4
PRINCIPLES OF TOXICOLOGY

A few years ago the terms "toxic" and "hazardous" entered the American vernacular alongside such words as "apple pie" and "baseball." Nowhere does science clash more directly with popular perception than in the use of these adjectives; the scientific and popular understanding of these terms have relatively little in common. Ask any American with limited knowledge of toxicology for a definition of "toxic" and he or she will have no trouble providing one: "It's something that causes illness, makes you sick or kills you." Ask a toxicologist, by contrast, and he or she will tell you that "toxic" is a relative term that has little meaning without further qualifications: How much? What route of entry into the body? Over what period of time? These questions are but a few a toxicologist would ask.

TOXICITY AND DOSAGE

Even a chemical generally considered harmless, safe, or nontoxic in small doses can be toxic at some level. Oxygen, for example, essential for life and generally composing about 20% of our air, is quite toxic when breathed in 100% concentrations for a couple of days. In newborn infants (particularly small premature neonates), an overdose of oxygen can lead to a toxic condition known as retrolental fibroplasia, a vascular overgrowth involving the retina, which may produce blindness. Adults maintained on high levels of oxygen may develop oxygen intoxication, a lung disorder that is frequently fatal.

Even water can be toxic, not only when inhaled—which is obviously deadly—but also when consumed in large quantities. Too much water, say a couple of gallons, drunk too quickly can overwhelm the body's ability to cope with the fluid overload. The result is water intoxication characterized by brain swelling, coma, and, at times, death. Now we might ask, "Are oxygen and water toxic or are they nontoxic?" Clearly under customary circumstances they are nontoxic, but under other circumstances, they switch and become toxic. Toxicity then, is not a quality inherent in certain substances: it depends on the circumstances of exposure.

Consider medications. Two aspirin can relieve a headache. A full bottle is lethal. Most prescribed medications have precise dosage regimens within which they are generally safe. Deviations may produce toxicity. The amount of deviation necessary to produce toxicity is known in pharmacology as the therapeutic/toxicity ratio. The larger the ratio, the less likely the drug is to be toxic at therapeutic dosages. Penicillin, for example, is very minimally toxic (except in those people with allergies) unless huge dosages are given, well over the drug's customary regimen. Digitalis and lithium, by contrast, have small therapeutic/toxicity ratios. People develop toxicity at dosages quite close to the therapeutic dose and patients taking these drugs need to be monitored very closely. Thus, even drugs cannot be classified simply as toxic or nontoxic. Their relative toxicity can be defined; however, at some level, all drugs are toxic.

Ethyl alcohol, the kind in beer, wine, and hard liquor, is clearly toxic at some level. In fact, it is drunk mainly for its toxic effects—that of altering mood. But one teaspoonful won't likely make anybody drunk or produce any other toxic effects. Ethyl alcohol also illustrates another toxicological issue: that of acute versus chronic toxicity. Alcohol intoxication, or drunkenness, is an acute effect—something that happens immediately after drinking ethyl alcohol. Chronic toxicity, by contrast, may occur only after years of use. Liver disease, neurological degeneration, and heart muscle dysfunction are but a few of the long-term toxic effects of chronic alcohol use. But such effects only occur if chronic use reaches a certain level, a level that most would associate with alcoholism. One could drink a glass of beer everyday for fifty years and never develop any of these complications. The dosage is simply too small. Thus, chronic toxic effects of chemicals depend upon dosage, as well as duration and extent of exposure. Acute effects depend upon dosage and the inherent degree of toxicity alone. This fact—the dose makes the poison—is illustrated in Figure 4.1.

Industrial solvents such as xylene, toluene, and methyl ethyl ketone follow precisely the same rules, as do aspirin and ethyl alcohol. At some dosages their toxicity is minimal or nonexistent; at other dosages, it is not. While it is true that industrial chemicals are not intended for consumption, as drugs are, that fact alone has no bearing on their relative toxicities. Some drugs are far more toxic than some organic solvents and vice versa. However, although it takes far fewer grams of digitalis to kill a person than it does of trichloroethylene, the latter (an industrial solvent) is widely perceived to be more toxic than the medication digitalis. Chemicals that aren't meant to enter our systems are commonly considered more toxic or hazardous than those that are, even though in many cases this perception is completely inaccurate. Table 4.1 illustrates this comparative toxicity; the smaller the number, the greater the toxicity. Here are

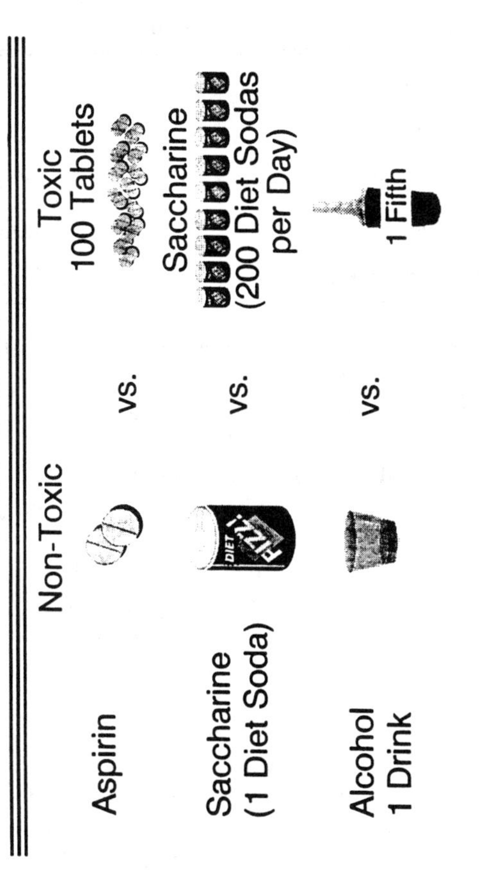
THE DOSE MAKES
THE POISON
Non-Toxic
Toxic
Aspirin
vs.
100 Tablets
Saccharine
(1 Diet Soda)
DIET
vs.
Saccharine
(200 Diet Sodas
per Day)
Alcohol
1 Drink
vs.
1 Fifth

Figure 4.1

listed approximate LD-50's: the amount required to kill one-half of a test animal population. This is only one measure of toxicity—a relatively crude one—but it does provide a relevant comparison. Note that table salt is more toxic (in terms of lethality) than is trichloroethylene.

Table 4.1 Approximate LD-50's

Chemical Agent	Dosage in mg/kg to Produce Lethality in 50% of Test Group
Nicotine	1
DDT	100
Chlordane	400
Table salt	3,000
Trichloroethylene	5,000

Like all chemicals that enter the human body, industrial chemicals are transported, metabolized, and eliminated. They also follow the same principles of dose response—some amounts are harmful, smaller quantities are not. It seems to me that this fact is commonly overlooked or consciously misstated. People have come to believe that any amount of a chemical with a frightening multisyllabic name in our air or water must be dangerous and should be eliminated. One can understand why this belief has grown. We ingest drugs or ethyl alcohol consciously. We are comfortable with the concept of dose-response. We are willing to accept some risk of toxicity in exchange for the benefits we derive from their use.

By contrast, we don't want a contaminant like trichloroethylene in our drinking water. Since there is no reason for it to be there, there is no perceived benefit. Therefore, the perceived risk looms as the focal issue, tempered by the fact that absolute safety can never be proved. In parts per billion dosages, it is most likely harmless, but factual concepts of dose-response are often outweighed by these emotional considerations. A typical dose-response—the relationship between dose and toxicity—curve is shown in Figure 4.2.

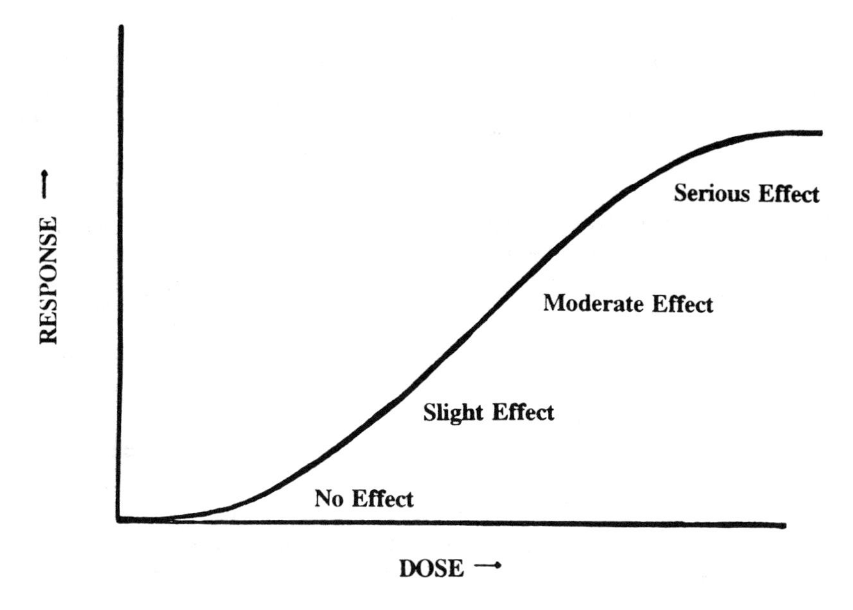

Figure 4.2 Dose-Response Relationship for a Typical Chemical

Here we see that at low dosage, there is no toxicological effect. As the dosage increases, the probability of toxicity increases. Dose relationships in toxicology have become extremely complex and sophisticated in recent years. Many of the effects besides death are evaluated in both human beings, when such data are available, or, more often, in experimental animals. Effects on each organ system or specific functional activities may be studied. The liver, brain, peripheral nerves, kidneys, heart, and lungs are among the common organ systems studied. Functions such as reproduction, development (studies to assess birth defects), cancer, and immunological alterations are also examined. To further complicate the matter, studies are performed in a variety of experimental animals of different species, sexes, and specific genetic variations. Once all these data are acquired and compiled, we have supplemented our dose-response curve with specific organ-system responses. These responses that are characteristics of specific test animals may not necessarily be those of human beings.

The following figures illustrate No Observed Adverse Effect Levels (NOAEL) and Lowest Observed Adverse Effect Levels (LOAEL) for trichloroethylene for various effects, in various animals, with either oral or inhalational exposure. Figure 4.3 illustrates various organ system effects, ranging from acute to chronic, in a variety of animal species. Figure 4.4 compares animals and humans.

The criticality of dose-response cannot be overstated. Not only does it exist, but it is relatively consistent (with some individual variation) and predictable from person to person. If that were not so, there would be no safe medications; two tablets might help one patient, but kill another.

There would also be no way to protect workers from exposure to chemicals. If one could not predict with some degree of accuracy how workers would respond to chemicals in the workplace, there could be no rational occupational standards. If there were no known safe levels, every house painter would have to wear self-contained breathing apparatuses, as would the homeowner after the paint job. The same precautions would be necessary for every lawncare worker, every printer, every employee in every dry-cleaning establishment, every floor finisher, every janitor, every gas station attendant, every mechanic, and every toll-booth attendant.

All of these people are exposed to chemicals that can be dangerous under certain circumstances at some levels, but that, if used with proper ventilation in adequate working conditions, are perfectly harmless. That does not mean that a painter or any of these others could not be exposed to unacceptably high levels. Overexposure can occur under conditions of poor ventilation in tight spaces, but generally does not. Consumers, too, can safely use products that, under some circumstances, can be hazardous. We use plastic glues, epoxies, and typewriter correction fluid comfortably without wearing respirators. Those same products are sometimes put into plastic bags, placed over the head, and deeply inhaled to produce a form of "high" or intoxication. These high dosages produce acute toxicity (the intended purpose of the exercise). They can also produce unintended side effects or overdosages—they can kill immediately or produce chronic toxicity including brain damage.

It would be ludicrous to suggest that we should all wear respirators when we assemble a model airplane, just because airplane glue can be toxic if put into a plastic bag and placed over the head. The same is true of household paints, furniture polishes, varnishes, various cleaning fluids, motor oil, and gasoline. We don't need to wear respirators when we use

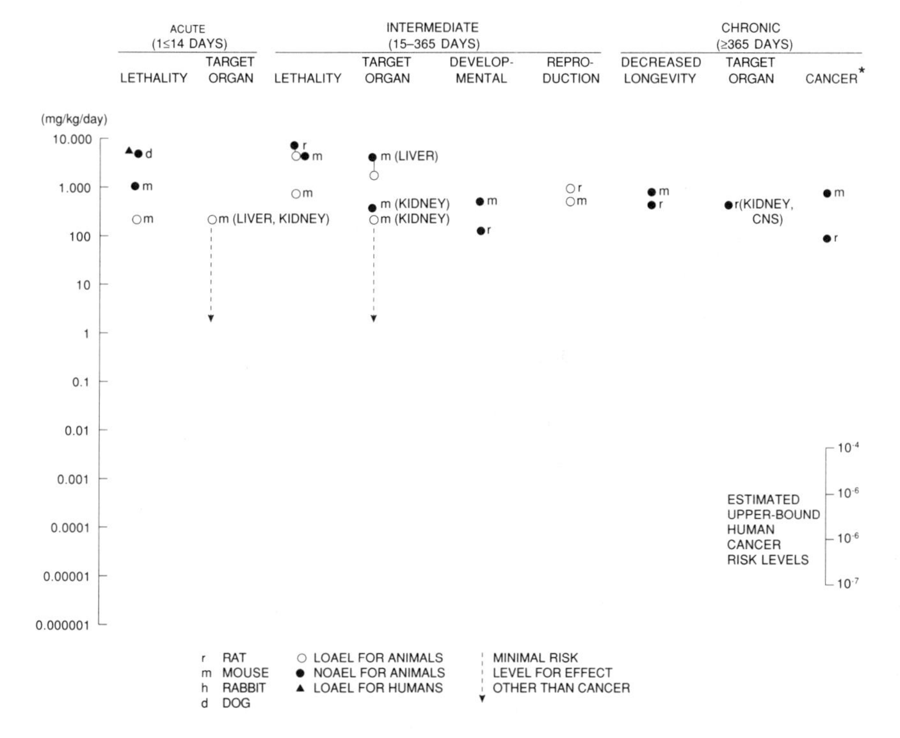

* Doses represent the lowest dosages tested that produced a tumorigenic response in the studies and do not imply the existence of a threshold for the cancer end point.

Figure 4.3 Levels of Significant Exposure for Trichloroethylene - Oral

Reproduced from Agency for Toxic Substances and Disease Registry, *Toxicological Profile for Trichloroethylene,* NTIS, Springfield, VA, 1990.

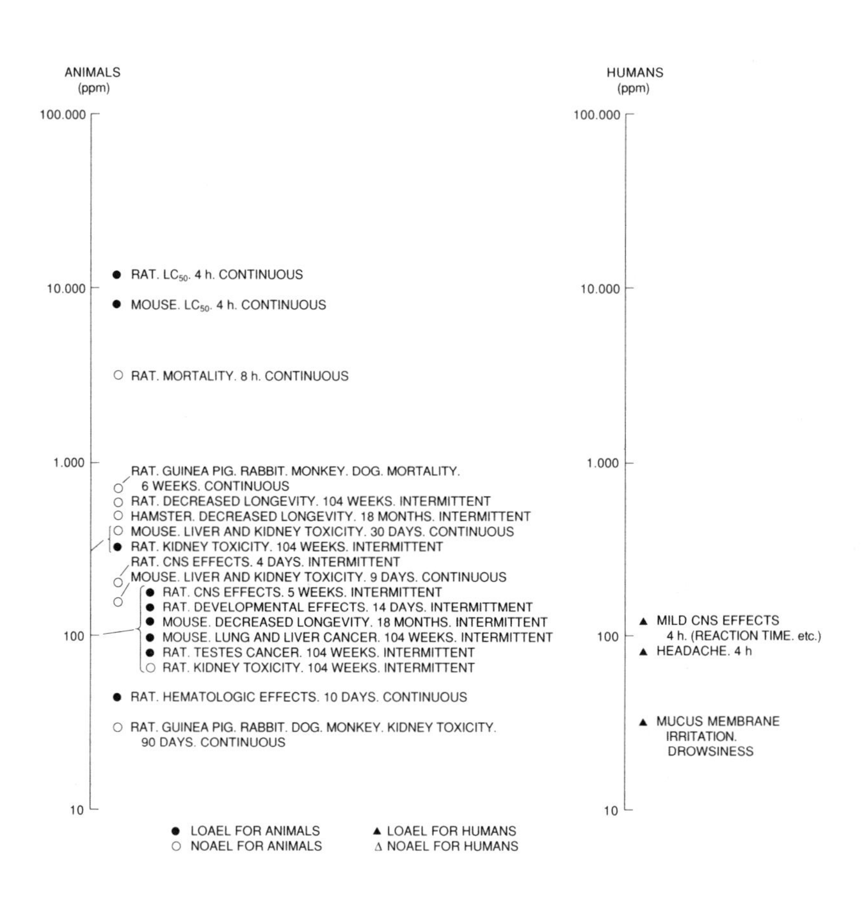

Figure 4.4 Levels of Significant Exposure for Trichloroethylene - Oral

Reproduced from Agency for Toxic Substances and Disease Registry, *Toxicological Profile for Trichloroethylene,* NTIS, Springfield, VA, 1990.

these materials, but we do need to minimize our exposure (dosage) by maintaining adequate ventilation.

DOSE-RESPONSE TERMINOLOGY

Over the past 10 years new terminology has been established to refer more completely to these expanded data. The terms NOAEL and LOAEL are among the many relatively new ways of expressing this mass of experimental observations. Some of these more common terms are delineated and defined in Table 4.2. These numbers vary not only according to which specific animals were used in the study, but also depending upon the organ system or functional effect being examined and upon the route of exposure. The fact remains, however, that although these dose-response relationships may be complicated and varied, they invariably exist. For almost all toxic effects, there are levels below which we see no toxic effects under any experimental conditions. The one exception is in true allergic responses in which the susceptible individual may respond to extremely low dosages. True allergy has distinct, reproducible characteristics. Relatively few chemical substances are known to be allergens.

SERIOUS VERSUS BENIGN EFFECTS

The term "toxicity" does not denote severity. A toxic effect may be very serious, even fatal, or it can be minor and benign. The irritant effect of an onion, which causes burning and watering eyes, is a form of mild toxicity. The light-headedness associated with the consumption of one beer is also a form of toxicity. The same is true of a bee sting, poison ivy, a rash on the hands from detergents, an upset stomach from aspirin use, or a scratchy throat caused by household ammonia. Toxicity in this sense is an everyday experience: an unimportant minor annoyance. "Toxicity" could refer to the mass fatalities that occurred in Bohpal, India, from a large release of methyl isocyanate; the birth defects that were produced by the drug thalidomide; and the deaths of children who eat arsenic-containing rat poisons. Thus, toxicity runs the gamut from quite insignificant to very serious.

It is likely that each person who uses or hears the word "toxic" has his or her own definition. Many might think of death or cancer as the endpoint of toxicity. Miscommunication of this term can create serious differences in perceptions. For example, a school building may have some problems with indoor air. An engineering group might perform some measurements and write a report which speaks of "toxic" levels of formaldehyde. In actuality, the levels found (0.1-0.2 ppm) might produce some irritant effects—purely speaking, a form of toxicity—yet they would not cause serious harm or major risk. Yet the parents of the children, hearing the report describing "toxicity," might interpret that as suggesting a life-threatening hazard for their children.

Table 4.2
Commonly Used Toxicity Terminology and Its Meaning

FEL	**Frank Effect Level** - Exposure level that produces a significant increase in unmistakable adverse effects.
IDLH	**Immediately Dangerous to Life or Health** - The maximum level of an environmental contaminant from which one could escape in 30 minutes without impairing symptoms or irreversible health effects.
LCLO	**Lethal Concentration (LO)** - The lowest concentration of a chemical reported to have caused death in animals or humans.
LC50	**Lethal Concentration (50)** - A calculated concentration of a chemical in air to which exposure for a specific length of time is expected to cause death in 50% of a defined experimental animal population.
LD	**Lethal Dose**
LD50	**Lethal Dose (50)**
LOAEL	**Lowest Observed Adverse Effect Level**
MRL	**Minimal Risk Level**
NOAEL	**No Observed Adverse Effect Level**

NOAEL stands for no observed adverse effect level, the dosage that produces no effect in any of the test animals studied. The LOAEL is the lowest observed adverse effect level, the lowest dosage that produces an observed effect in any of the experimental studies. An example of how LOAELs and NOAELs are commonly displayed is shown in Figures 4.3 and 4.4.

Thus, it is important to remember that, by themselves, the terms "toxic" or "toxic effect" say nothing about severity. These terms are similar to the phrase "being sick," which could refer to a cold, or to a headache, or to terminal cancer.

ABSORPTION

Some chemicals may cause direct irritation, burning or allergic reaction involving the skin or mucous membranes lining the nose or throat as a result of direct surface contact. Short of such surface contact effects, toxins produce toxicity only after they are absorbed into the body. We do not develop drug toxicity by walking into a pharmacy even though a great many dangerous and potentially quite toxic substances are present there. They don't enter our bodies just because we walk into the room where these substances are. The mere existence of a chemical waste disposal site in an area does not mean that the chemicals from that site can produce toxicity in people who live nearby. The chemicals have to leak out of the site and be absorbed by the people to cause toxicity.

Chemicals can be absorbed into us in three ways. They can be swallowed (ingested). They can be breathed (inhaled). Or, they can be absorbed through the skin itself. Chemicals in the workplace most commonly are inhaled and, somewhat more likely, absorbed through skin contact. Chemicals have variable volatility, meaning they may or may not leak into the air. Only when they do, can they be inhaled. By the same token, some jobs may result in extensive skin contact with chemicals that may or may not be readily absorbed. It depends upon the properties of the particular chemical. These modes of absorption are shown in Table 4.3.

Ingestion in the workplace is rare, but it can happen accidentally. Siphoning gasoline by sucking is a potential hazard. Another is eating without washing contaminated hands. This can happen to farm workers, who may thereby become acutely intoxicated by pesticides.

Table 4.3 Modes of Chemical Absorption into the System
Inhalation
Ingestion
Skin Absorption

Environmental sources of potential toxins include air, water, food, and soil. Drinking water may contain contaminants—city air always does. Children may eat dirt contaminated with lead. Pesticides or other residues are often found in the foods we eat. For the most part, such contamination is at levels too low to produce any acute effects. Exceptions might be an acute, major release from a chemical plant such the one in Bohpal, India, an overturned tank car carrying chlorine gas, or the accidental ingestion of lead-contaminated oil or chips of lead-based paint.

How much of a particular chemical is absorbed by each route depends upon a variety of factors including the concentration of the chemical, the duration of the exposure, and certain physical and chemical properties of the chemical. For a chemical to produce systemic toxicity when swallowed, it must be absorbable through the intestinal tract. Not all chemicals are. For example, dioxins are tightly bound to certain dirts making absorption through the ingestion of dirt unlikely.

Skin absorption, too, is highly dependent upon the nature of the chemical. Some chemicals will never be absorbed through the skin; others are quickly taken up. The new application of medications such as scopolamine through dermal patches takes advantage of individual chemical rates of absorption. A patch can be worn for several days and provide gradual, continuous absorption of the drug into the bloodstream.

DISTRIBUTION, METABOLISM, AND ELIMINATION

Half-life, Distribution, Storage, and Pharmaco— and Toxicokinetics

Once a chemical enters the body, many things happen to it. It is distributed by the bloodstream to various sites. Which site depends upon the chemical's physiochemical characteristics. Some chemicals are readily transported to the brain—alcohol, for example—while others have a more difficult time getting there. Some chemicals, like dioxins and PCBs, are highly soluble in fat and are taken up by fat cells. Others, such as lead and other metals, are taken up primarily by the bones. Where a particular chemical moves and how it is apportioned among various bodily tissues is known as "distribution."

Chemicals are handled by the body in different ways—and, even for the same chemical, many different events may take place. Ethyl alcohol is partially metabolized (chemically altered) in the liver and partially exhaled in the breath unchanged. Typically, chemical conversions make chemicals less toxic and easier for the body to eliminate. They may, for example, be made more water soluble, permitting excretion in the urine. There are, however, exceptions. Sometimes chemicals can be converted by the body to more hazardous forms. Potential carcinogenicity of some chemicals may come about through such changes, that is, formation of epoxide intermediates.

Thus, the absorption, distribution, and metabolic fate of chemicals can be quite complicated. An additional complicating factor is the species differences among laboratory animals and between animals and human beings.

TOXICITY: A FUNCTION OF CONTEXT

Now we return to the original question. Are given substances toxic or nontoxic? Are glue, paint, gasoline, typewriter correction fluid, floor waxes, and varnishes toxic or not? The answer is neither and both. To most people the term "toxic" is an absolute term, conveying a meaning peculiar to each individual and his or her own perception. Reductionist thinking of this sort simplifies and distorts communication. Newspapers can write headlines describing a "toxic chemical spill" or "the most toxic chemical known to humans," and everyone will know what that means, even though everyone's interpretation will differ and few people will interpret accurately.

It is important for people to begin to understand the relative nature of the term "toxicity" so that they may put it into context and make rational judgments. These examples from everyday experience illustrate the fact that the term "toxic" is not an adjective that can be affixed without qualification to a chemical. It is not an objective term like "red," "wooden," or "alive." Rather, it is a subjective term like "big," "attractive," or "slow." A house may be big when compared with a breadbox, but it's small when compared with a skyscraper. A train may be slow when compared with an airplane, but it is far faster than an ant. And a chemical is only relatively toxic by comparison with something else and, most important, as a function of the dosage and the route of administration. Context is all important in the assessment of toxicity.

HOW CHEMICALS PRODUCE ADVERSE EFFECTS

The bite of a coral snake paralyzes the muscles, stops one from breathing, and causes death quite promptly. This effect is unique to this snake's venom. By contrast, the king cobra also produces a deadly bite, but death is caused by an entirely different process. The toxin causes the blood to lose its clotting capability and the person bleeds to death. Cyanide attaches to specific enzymes within cells that are responsible for intracellular respiration and for the production of energy. Cyanide prevents cells from using oxygen, in effect, producing suffocation. Carbon monoxide attaches to hemoglobin, the protein in the red blood cell that carries oxygen. It attaches far more strongly than oxygen, thereby displacing it and making it unavailable for oxygenating cells. In effect, it, too, produces suffocation, but by a mechanism quite different from that of cyanide.

Chemicals commonly exert toxic effects through specific and characteristic processes. They may interfere with the transmission of nerve impulses; they may affect blood coagulation; they may interfere with cellular respiration; or they may block the uptake or use of certain nutrients.

Organophosphates, a class of commonly used pesticides, act by mimicking a normal body chemical, acetylcholine. They do this by attaching in a sort of lock-and-key fit to the enzyme, acetylcholinesterase, that is designed to break down that chemical. That attachment prevents the enzyme from acting, which results in the excessive accumulation of acetylcholine. Acetylcholine acts as a transmitter of certain specific nerve messages. When it is enhanced, all of its messages are amplified. Its normal effects include constriction of the pupils, sweating, salivation, fluid secretion in the intestinal tract, slowing of the heart rate, and others. Organophosphate poisoning causes all of these effects.

TARGET ORGANS

Typically, toxicity associated with chemicals and other toxic agents involves certain patterns and certain affected organs. To be sure, the effects of some chemicals are more pervasive than the effects of others, but even when effects are widespread they are often consistent. For example, upon inhalation into the body, asbestos lodges in the lungs. The lungs are the sole site of its toxic effects. Organophosphates produce widespread effects, but all are related to the accumulation of acetylcholine. Thus, all of its effects, though they involve the systems ranging from the eye to the intestinal tract, are nonetheless invariably consistent. Lead affects the nervous system, the blood-forming system, and the kidneys, but it has, by contrast, little effect upon the liver. Some chemicals predominately affect the liver, but have little effect upon the kidneys.

Benzene may produce aplastic anemia or even leukemia—both effects are on the blood-forming system—but it has little kidney or liver toxicity.

Toxic specificity is quite important. It gives us some clue as to how toxicity occurs and/or where the agent normally goes in the body. More important, it tells us what to look for. In the case of medications, knowledge of these effects provides a protocol for patient management. We know to check the blood count, liver function, or kidney function, for example. Similarly, this knowledge and specificity helps to guide the monitoring of workers exposed to potentially hazardous chemicals. Depending upon which chemicals they are exposed to, we can tailor a program to hone in on those parameters of bodily function most likely to be adversely affected by their exposures.

This concept of specificity of chemical effect is not commonly understood and is, I believe, becoming lost in the turmoil of chemical misperceptions. It seems more and more common for people to think that exposure of any kind to any chemical can produce any and all sorts of disorders.

INDIVIDUAL SUSCEPTIBILITY

A growing question is whether the toxicological characteristics of chemicals are inherent in individual chemicals or whether their effects vary from person to person. With certain notable exceptions, toxicological effects have generally proven to be characteristic of the chemical. For the most part, people react similarly to the same substances. If that were not the case, there could be no medications, because one would never know what a particular drug might do. One patient might develop seizures; another, a heart attack; a third, kidney failure, and a fourth, the intended beneficial effects, all at the same dose.

The same is true of industrial chemicals. Workers do not have enormous differences in susceptibility to the adverse effects of most chemicals. There are certainly some differences in individual response, but generally within a range of dosages or exposures.

The one clear exception is the case of true allergy. Allergy is an individual response. It means that the person develops antibodies or other immunological factors that interact with the agent. While one person may be allergic, another will not be. Some chemicals can produce allergies in susceptible individuals, while many do not.

One must be very careful in these days of chemical misperceptions not to overcatalogue perceived chemical effects. Many of these are not chemical effects at all. We discussed in Chapter 2 the clinical ecologists and the concept of "multiple chemical sensitivities." Without clear and reproducible chemical characteristics, it is dangerous and scientifically erroneous to hypothesize a mechanism derived from "individual susceptibility" to explain perceived adverse consequences of chemical exposures.

Thus, to date, except for instances of true allergy, individuals generally respond similarly to similar chemicals. This may be further specified in coming years as great research attention is currently being devoted to this issue, but for now, it remains a good rule of thumb.

TYPES OF TOXICITY

Thus far we have been speaking largely of acute toxicity—effects that occur promptly, seconds to at most a day or two after an exposure. Certain other effects may be delayed. Depending on the chemical and the degree of exposure, effects can appear any time between the instant of contact with an agent and several decades later. An instantaneous toxic reaction is called "acute toxicity." For purposes of this discussion, acute and chronic toxicity are sufficient for our needs.

Dramatic, sudden poisoning, such as Socrates and the hemlock, a prick from a poison dart, the bite of a rattlesnake, and the release of methyl isocyanate over Bohpal, India, are all examples of acute toxicity. The toxin entered the victims' bodies (in these cases by the different routes of ingestion, injection, and inhalation) and they became sick or died almost instantly. A substance exerts an immediate life-threatening effect by interfering with an essential life support function—generally circulatory, respiratory, or neurological. Nerve gases interrupt nerve impulses, which interferes with breathing. Methyl isocyanate produces a severe lung reaction in which fluid floods the tissue, thus preventing oxygen from moving in and out of the air cells. Acute toxic effects may also be benign. Ammonia used as a simple household cleaner will irritate the eyes and nose while it is being used. This is an acute toxic effect, but it is one that is generally not serious and abates quickly after the exposure ends.

Chronic effects, by contrast, develop only after months or years of exposure. Most often they occur after long periods of repetitive low-level exposures, but they can occur after a large one-time exposure. Asbestosis is an excellent example. Asbestosis is a disease in which the lung tissue is scarred because it responds to the presence of asbestos fibers. Such scarring begins to occur at the time of the initial exposure, but it does not produce a clinical problem until it becomes quite extensive. This may take many years of ongoing and intense exposure. In fact, asbestosis doesn't occur unless the individual is heavily exposed for a long period of time. It is a chronic disease, requiring chronic exposure.

Lead poisoning, too, is commonly a chronic illness. One can drink large amounts of a lead-containing fluid and develop acute lead poisoning, but the lead poisoning we most commonly see is chronic. It occurs in children who, over a period of time, have eaten lead paint and/or dirt containing lead; it comes from air contamination, by lead-containing gasoline (although less frequently now that most gasoline is lead-free); and it comes from lead pipes contaminating the water that flows through them. Lead produces a chronic disease because it accumulates in the body. Most of the lead that is ingested is absorbed by the bones in which it is stored. The bones serve as a reservoir, releasing steady amounts into the bloodstream. Once the reservoir is heavily loaded, the blood level remains high and various organs become affected. Thus, lead poisoning commonly is a chronic, rather than an acute, form of toxicity.

CHRONIC TOXICITY: THE LIMITATIONS OF DATA

When it comes to industrial chemicals, we know much more about acute toxicity than we do about chronic toxicity. Acute toxic effects are often dramatic and always immediate. This makes them easy to relate to, say, an occupational exposure. If a worker enters a tank to clean it, loses consciousness and has to be carried out, acute toxicity is apparent. From that point on, that facility, it is hoped, will not permit anyone to enter such a tank without proper protective equipment. Chronic effects, by contrast, are often far harder to relate to particular exposures.

In the case of asbestosis, workers would become ill many years after leaving the work site. It took clever investigators to trace the disease to the earlier exposure. It is easier to relate chronic exposures to chronic toxicity when there are footprints that particularly characterize the illness. Asbestosis, in its classic form, is a relatively unique disease, not readily confused with other diseases. In less severe cases, however, even today, the diagnosis may be uncertain. This is one of the reasons that there are so many fights over compensation for asbestos-related disease. The worker may be convinced it's asbestos related. The evidence, however, may be far less conclusive. Lead poisoning, when severe, is quite readily identifiable and is accompanied by high blood lead levels and the other biological markers that characterize it. There currently rages a debate, however, over the more subtle effects of chronic lower-level lead exposure.[1-4] Some believe that reactions can be caused by lower levels; others argue that the presumed effects may not have anything to do with lead. The issue of subtle effects is quite integral to our study of the line between the known and the possible. The lead issue will be discussed more extensively later in Chapter 15. Here it is relevant to point out the difficulties in resolving chronic toxicity questions, particularly with low environmentally-acquired dosages and when the effects that are measured are subtle.

Lead, at high levels, produces a rather unique set of disorders. These include a type of anemia, a characteristic problem with the nerves to the extremities, and a typical brain dysfunction. Severe lead poisoning is recognized clinically by these features and is readily confirmed because blood lead levels are high and the bone lead is visible on X-ray. Most first-year medical students can recognize and diagnose a case of full-blown lead poisoning. In less severe cases of this disease, by contrast, the characteristic features begin to trail off and the distinct diagnostic qualities become fuzzier.

At the most subtle extreme, the diagnosis becomes tenuous and the very existence of the disease uncertain. It is at this extreme that much of the debate about the effect of low levels of lead in the air upon children (with its regulatory implications) takes place. It is at this end that the truths trail off and regulatory science appears. Discussions of this decade have centered upon the effects of low levels of lead upon the mental development of children. Again, when levels are high, there is little argument—lead can affect intellectual function. But the more subtle claimed effects of lead, such as minor alterations in intelligence quotient (IQ), are less characteristic and far more difficult to link conclusively to lead and lead alone.

Unlike the clear fingerprints of full-blown lead poisoning, functional alterations found in cases of low-level poisoning are of a minor nature and, more important, are derived from many possible sources. A drop of four IQ points is small; indeed, such factors as artifacts of test taking, not to mention environmental, dietary, and genetic influences, could easily have produced such minor changes. Currently, much of the regulatory battles over the chronic effects of lead relate to such subtle or uncertain effects.

Another recent example of fuzzy science is the investigation of chronic brain effects of a variety of organic solvents. We know that large amounts of certain organic solvents can produce long-term and significant brain injury. We see this, for example, in glue sniffers who, after years of getting high from the toluene in glue, develop serious and clear-cut behavioral disorders associated with actual "holes" in the brain that can be seen with various forms of brain imaging.

More recently, investigators have questioned the effects of lower-level occupational exposures to a variety of organic solvents—most notably in painters and printers—during customary occupational use. In numerous studies, tests of mental performance and coordination have suggested that chronic exposure to certain solvents may lead to varying degrees of dysfunction over long periods of time.[5-10] More recently, some studies have purported to suggest that even rather low levels of exposure may produce similar alterations.[11-12] Others, by contrast, have found the opposite to be true.[13]

These assessments of chronic toxicity suffer from many of the scientific methodological problems noted in Chapter 3. In most cases, actual exposure levels are unknown. Thus, the certitude that one is actually comparing like situations as a potential cause, examined in the context of similar cases, is not present. Also, the functional tests are not as clear-cut as a blood measurement. Rather, they depend upon mental effort and the interpretation of the product of that effort. They can be influenced by many factors, both intangible and unknown, such as interest, nervousness, ability to concentrate, use of medications, and tiredness. Thus, the results are quite unpredictable, rarely measuring with any specificity chemical toxicity.

Once more, those features that are not specific fingerprints of the disorder raise significant questions about the cause. Here, again, we have a matter of possible chronic toxicity due to low-level chemical exposure that has eluded effective research. The results of such exposure led to interpretations about chronic toxicity, which may or may not be correct. They are far less clear-cut from the scientific standpoint than our understanding of full-blown lead poisoning or asbestosis.

That does not mean that we should not attempt to study more subtle chronic effects of chemical agents upon people. I am concerned, however, that popular pressure for such answers is leading increasingly to hasty and poor science that commingles, eventually indistinguishably, with quality science, until every observation and report is accorded equal weight regardless of the quality of the data. It was exactly that practice that has led to false notions of causes of diseases in the past such as the autointoxication theory and the more recent popular concept of chronic yeast infection syndrome.

Information about chronic toxicity is, today, the prime mover of the regulatory process. Thus, understanding the quality of chronic toxicity data is essential to monitoring the quality of the science that drives regulation. The preeminence of chronic toxicity concerns in regulation has come about for a very simple reason. Most, but not necessarily all, issues of acute toxicity from environmental chemicals have been addressed. In the course of the routine American working day, workers are now protected from acute toxic hazards. Accidents still occur, of course; a respirator may fail or a major spill can take place. But a worker generally will not clean the inside of a chemically contaminated tank car without wearing protective equipment. Occupational settings in which high levels of mercury or arsenic are present will customarily be controlled in the 1990s. The same is true for environmental contamination or pollution. Companies are now rarely dumping potentially toxic materials directly into streams or lakes. Nor are they commonly discharging acutely dangerous concentrations of chemicals directly into the air through smokestacks. We will speak more in later chapters about the agencies and acts that have controlled such practices. Today acute toxicity is now averted through appropriate industrial control practices.

Having achieved control over acute toxicity, the public, the legislatures and the regulatory agencies have turned their attention to chronic exposures and long-term effects. It is this arena in which the pressure for action is most intense. Paradoxically, it is also this arena in which our science is the weakest and the studies most difficult to carry out.

CARCINOGENICITY AND MUTAGENICITY

Health-related regulation centers largely upon chemicals as potential carcinogens. So, too, do popular concerns and perceptions. Consequently, more will be said about chemicals and cancer in subsequent chapters when we deal with both regulation and perceptions. The science of carcinogenesis is quite complicated; what follows is, therefore, only a brief survey.

The development of cancer is a multifactorial process beginning with changes in the gene of a cell. This change programs a cell to reproduce in an uncontrolled fashion in the process called neoplasia or malignancy. Numerous mechanisms, some known, many unknown, can bring about such changes. They may arise spontaneously. They may occur through activation of certain sections of the gene known as "oncogenes," which are preprogrammed to produce a cancer. They may occur because of physical factors such as exposure to radiation. They may be produced by dietary factors. They may come about due to infectious causes such as viruses, or they may be caused by chemical carcinogens.

When any of these things act upon the gene to start the malignant process, the action is known as "initiation," because it initiates, or begins, the cancer. It is also known as a deoxyribonucleic acid (DNA)—reactive carcinogenic process or a genotoxic process because it acts directly on the gene. It is believed that such gene alterations occur with regularity. We have many reparative mechanisms that are used to "fix" such damaged genes. And even if a genetic alteration does occur and persist, many additional events must take place before the cancer is actually manifested. Thus, the first stage is initiation, but it is only one link in a complex chain of events. Initiation alone does not cause cancer.

In Pitot's classification (Table 4.4), there are at least two major stages required for the actual expression of a malignancy.[14] After initiation, the second stage, commonly known as "promotion," that is an epigenetic phenomenon, meaning that it occurs outside the gene itself. There are thought to be many types of promotional processes which permit the damaged DNA to be expressed in the form of cancer growth.

The third phase of cancer growth is "progression." Here the promoted tumor is helped along by the agent or environment causing the progression.

Some chemical carcinogens are genotoxic, initiating the process. The most notable are, paradoxically, the alkylating agents, which are drugs used to treat cancers. They can act directly upon the DNA, causing it to mutate. Certain industrial chemicals have also been shown to be genotoxic. These include certain alkene epoxides, active halogen compounds (i.e., bis(chloromethyl)ether), polycyclic aromatic hydrocarbons, nitrosamines, and others. Many other chemicals act primarily through epigenetic mechanisms, producing either promotion or progression. Saccharine and dioxins fall into this category.

Table 4.4 Pitot's Classification of Stages in Carcinogenic Expression

Initiation	Promotion	Progression
Irreversible with constant"stem cell" potential. Initiated "stem cell" not morphologically identifiable	Reversible	Irreversible Measurable and/or morphologically discernible alteration in cell genome's structure
Efficacy sensitive to xenobiotic and other chemical factors	Promoted cell population existence dependent on continued administration of the promoting agent	
Spontaneous (fortuitous) occurrence of initiated cells can be quantitated	Efficiency sensitive to dietary and hormonal factor	Growth of altered cells sensitive to environmental factors during early phase
Requires cell division for "fixation"		
Dose response does not exhibit a readily measurable threshold	Dose response exhibits measurable threshold and maximal effect dependent on dose of initiating agent	Benign and/or malignant neoplasms characteristically seen
Relative effect of initiators depends on quantitation of focal lesions follo wing defined period of promotion	Relative effectiveness depends on time and dose rate to reach maximal effect and dose rate	"Progressor" agents act to advance promoted cells into this stage but may not be initiating agents

Adapted from Multistage carcinogenesis: the phenomenon underlying the theories, in *Theories of Carcinogenesis,* Iverson, O.H., Ed., Hemisphere Publishing Company, Washington, D.C., 1988.

Some chemicals are classified as direct-acting genotoxins, meaning that they produce genetic changes without requiring any chemical alteration of their chemical structures. Others are indirect acting, meaning that they must be modified by the body before they can produce genetic change. The former class of initiator—direct acting—is also a mutagen, meaning that it produces mutations. When bacteria are treated with direct-acting genotoxins, such as certain chemicals or radiation, they develop genetic changes that alter certain of their functions. These changes are called "mutations." Mutations are thought to arise from the same sort of genetic damage that produces initiation or cancer potential. Thus, mutational changes in bacteria in response to chemicals have been used to find mutagens that could potentially also be carcinogens. The problem is that the more chemicals we test, including many occurring naturally in foods, the more mutagens we find.

Currently, hundreds of chemicals are known to produce or spread the growth of cancers in experimental animals. To date, approximately thirty are generally recognized as doing so in human beings.

SUMMARY

The field of toxicology is a rigorous discipline with a strong scientific base. The bedrock is dose response. As Paracelsius said in the 16th century: "The dose makes the poison." Exposure must be sufficient in duration and degree to produce an adverse response. The chemical must be absorbed into the body, unless it merely causes a topical effect—irritation or an allergic rash. Organs are affected in generally recognized and characteristic ways. Individuals, in general, respond similarly to toxic agents. In the case of allergies, there are clear differences in individual susceptibilities, but this is only clearly so in the case of true allergies.

There is widespread misunderstanding about many of these basic principles of toxicology. People view the very term "toxic" as an all or nothing phenomenon, rather than the broad gradient it actually represents. They fail to distinguish between degrees and classes of toxicity, which may range from trivial—minor irritation—to catastrophic—death. There is also the pervasive belief that industrial or man-made chemicals, somehow, as a class, are worse or more toxic than either naturally occurring chemicals or medicines. These misimpressions contribute to misdirected worries and regulatory responses based more upon emotions than upon established truths.

REFERENCES

1. Rutter, M., Raised lead levels and impaired cognitive/behavioral functioning: a review of the evidence, *Dev. Med. Child Neurol. [Suppl.]*, 22, 1, 1980.

2. Harvey, P.G., Lead and children's health-recent research and future questions, *J. Child. Psychol. Psychiatry*, 25, 517, 1984.

3. Ernhart, C.B., Lead levels and confounding variables, *Am. J. Psychiatry*, 139, 1524,

4. Smith, M., Recent work on low level lead exposure and its impact on behavior, intelligence, and learning: a review, *J. Am. Child Psychiatry*, 24, 24, 1985.

5. Cherry, N., Hutchins, H., Pace, T. and Waldron, H., Neurobehavioral effects of repeated occupational exposure to toluene and paint solvents, *Br. J. Ind. Med.*, 42, 291, 1985.

6. Eloffson, S., Gamberale, F., Hindmarsh, T., Iregren, A., et al., Exposure to organic solvents. A cross-sectional epidemiological investigation on occupationally exposed car and industrial spray painters with special reference to the nervous system, *Scand. J. Work Environ. Health*, 6, 239, 1980.

7. Gregerson, P., Angelso, B., Nielsin, T., et al., Neurotoxic effects of organic solvents in exposed workers, *Scand. J. Work Environ. Health*, 2, 240, 1976.

8. Seppalainen, A., Lindstrom, K. and Martelin, T., Neurophysiological and psychological picture of solvent poisoning, *Am. J. Ind. Med.*, 1, 31, 1980.

9. Eskenazi, B. and Maizlish, N.A., Effects of occupational exposure to chemicals on neurobehavioral functioning, in *Medical Neuropsychology: The Impact of Disease on Behavior*, Tarter, R.E., Van Thiel, D.H., and Edwards, K.L., Eds., Plenum, New York, 1988.

10. Hanninen, H., Eskelinen, L., and Husman, K., Behavioral effects of long-term exposure to a mixture of organic solvents. *Scand. J. Work Environ. Health*, 4, 19, 1978.

11. Morrow, L.A., Ryan, C.M., Goldstein, G. and Hodgson, M.J., A distinct pattern of personality disturbances following exposure to mixtures of organic solvents. *J. Occup. Med.* 31, 743-746, 1989.

12. Morrow, L.A., Ryan, C.M., Hodgson, M.J. and Robin, N. Alternatives in cognitive and psychological functioning after organic solvent exposure. *J. Occup. Med.*, 32, 444, 1990.

13. Otto, D., Molhave, L., Rose, G., Hudnell, H.K. and House, D. Neurobehavioral and sensory irritant effects of controlled exposure to a complex mixture of volatile organic compounds. *Neurotoxicol. Teratol.*, 12, 649, 1990.

14. Pitot, H.C., Beer, D. and Hendrich, S., Multistage carcinogenesis: the phenomenon underlying the theories, in *Theories of Carcinogenesis*, Iverson, O.H., Ed., Hemisphere Publishing Company, Washington, DC, 1988.

GENERAL REFERENCES

Kamrin, M.A., *Toxicology: A Primer on Toxicology Principles and Applications*, Lewis Publishers, Chelsea, MI, 1988.

Klaassen, C.D., Amdur, M. and Doull, J., *Casarett and Doull's Toxicology: The Basic Science of Poisons*, 4th Edition, Macmillan Publishing Co., New York, 1991.

Lewis, Richard J. *Hazardous Chemicals Desk Reference,* 2nd edition, Van Nostrand Reinhold, New York, 1991.

Ottoboni, A.M., *The Dose Makes the Poison*, Vincente Books, Berkeley, 1984.

Principles and Methods of Toxicology, 2nd Edition, Hayes, W.A., Ed., Raven Press, New York, 1989.

5
EPIDEMIOLOGY: ITS APPLICATION TO ENVIRONMENTAL TOXICOLOGY

THE DEFINITION AND PURPOSE OF EPIDEMIOLOGY

Epidemiology was, historically, the study of epidemics. We discussed in Chapter 2 John Snow's identification of the source of contaminated water, which led to eradication of cholera epidemics in London. At the turn of the century, Mary Mallon, a cook in Ithaca, New York, infected 50 or more people with *Salmonella typhi*, the bacterium that produces typhoid fever. At least five people died. Typhoid Mary, as she was later known, was discovered through the sleuthing of epidemiology.[1]

In the latter half of this century, epidemiology has moved beyond epidemics and infectious diseases. Its techniques have been adapted to the identification of a variety of sources of diseases in the population: dietary, lifestyle, chemically induced, and others. Now epidemiology more broadly includes the identification of causes and prevention of a wide variety of health problems. The formal definition of epidemiology has been expanded to embrace this broader role:

> Epidemiology is the study of the distribution and determinants of health-related states and events in specified populations and the application of this study to the control of health problems.[2]

The central characteristic of modern epidemiology, then, is that it evaluates distributions and causes of illnesses in human populations.

THE NATURE AND TYPES OF EPIDEMIOLOGICAL STUDIES

Epidemiological studies may study events that have already occurred, relationships that already exist, or future relationships that may be affected through experimental manipulation. Because of the very breadth of such studies, epidemiological methodologies are quite varied and diverse in nature.

The various types of epidemiological investigations may be divided broadly into two categories: descriptive and analytical, as shown in Table 5.1. The former, as the name implies, describes occurrences and relationships as they are.

Table 5.1 Types of Epidemiological Studies

Descriptive Studies	Analytical Studies
1. Population studies	1. Interventional studies
2. Case reports	2. Case-control studies
3. Case series	3. Cohort studies
4. Cross-sectional surveys	

DESCRIPTIVE STUDIES

Population Studies

One type of descriptive study is a population study, also called an ecological study. It compares one population with another and correlates observations. Because it makes these correlations, it is also called a "correlational study." For example, one may compare one population with another with respect to a disease incidence. Comparisons of the incidence of stomach cancer in the United States versus Japan is an example. Japanese living in Japan have a far higher incidence of stomach cancer than do people living in the United States. This type of descriptive, population correlational study tells us that there is some difference between these two groups. What that difference is and what the cause of the difference is cannot be assessed through this kind of study. We might hypothesize that diet is the cause, but it might also be a genetic difference, smoking, air pollutants, an infectious disease, or other unknown factors peculiar to one of the groups that accounts for the difference in disease incidence.

We learned in Chapter 3 the characteristics of experimental design. The reason that these correlational studies cannot pinpoint causation lies in their design. They are not designed in a way that permits examination of the independent variable; moreover, they are not controlled. They are merely observations, not scientific studies in the true sense.

Many current environmental epidemiological studies are actually population studies. Sometimes investigators forget (and the public almost always does) that these studies cannot establish causation. The misperception that such population studies tell us more about the cause of a health problem than they actually do has been called the "ecological fallacy" by Morgenstern.[3] Although these population studies cannot pinpoint causes, they are useful because they identify areas for further investigation. Thus, if we could find out in an analytical study why the residents of Japan have a higher incidence of stomach cancer, we might be able to intervene to prevent stomach cancer.

Case Reports/Case Series

Case reports and case series are forms of descriptive epidemiological studies. In a case report, a relationship between an outcome and a potential cause is observed. For example, a doctor may see a patient with a neurological disorder who is also taking a certain drug or was exposed to a particular chemical. He or she may decide that the relationship between the drug or the chemical and the disorder may be important and may write a case report describing his or her observation.

Such a description is merely one person's observation of correlated events—the drug or chemical and the disorder. It does not establish causation, for, again, it has none of the elements of experimental design. Many such case reports may be collected into a case series, which may add further strength to the association, but it still does not establish causation. Of the many thousands of such case reports and case series found in the medical and scientific literature, only a small fraction are ever shown through well-designed studies to be connected causally. Most fall by the wayside as interesting but irrelevant observations. Sometimes, however, these observations are significant and can lead to findings important to the public health. The now established causal relationships between tampons and toxic shock syndrome, thalidomide and birth defects, and diethylstilbestrol (DES) and vaginal cancers all began with case observations. However, causal relationships were not established until well-controlled, carefully designed, analytical studies were performed.

Cross-Sectional Surveys

Cross-sectional surveys are the last form of descriptive study. In these, the status of a population at some point in time is studied, both with respect to an outcome and to an exposure or other factor. For example, a questionnaire may ask all residents of a city about specific characteristics of health, of socioeconomic elements, of occupational activities, and of diet. These can be used by health officials to identify needs for public health programs, or they may be used to uncover relationships deserving further study. An example of a cross-sectional study would be measuring a factor, such as levels of vitamin C in the blood in a population with cancer. If vitamin C levels are either high or low, these findings may lead to further studies; however, they cannot establish causation for a number of reasons. For one thing, because both the vitamin levels and cancer are measured together, one cannot know from such a cross-sectional survey which came first. The cancer could have altered the vitamin C level, or the vitamin C level may have influenced the cancer.

ANALYTICAL STUDIES

Interventional Studies

The analytical studies of epidemiology are the ones that explore cause and effect questions most intensely and most meaningfully. These studies are designed to test the significance of relationships. Analytical studies may be experimental or observational. In the former, the characteristics of both the study group and the control group are set up in advance by the investigator. These analytical studies, known as "interventional studies" because the experimenter directly intervenes, are typical of those used in the testing of drugs prior to FDA approval. Two groups are matched. The control group is given a placebo drug; the experimental group is given the active drug. No individual knows to which group he or she belongs, nor does the observer (double blinding). The results are compared. This type of experimental, analytical, epidemiological study is the form of epidemiology that is closest to the basic sciences. By its very ability to control the variables and to set up the protocol, it is likely (if done properly) to provide meaningful results—to help establish or refute—causal relationships.

Case-Control and Cohort Studies

The other forms of analytical epidemiological studies are the case-control studies and the cohort studies. These are observational studies. These "softer" (less precise) epidemiological methods are far and away the most commonly employed to explore environmental and occupational toxicological questions. Unfortunately, they do not share the precision of the experimental/interventional studies because they rely not upon experimental control of events, but upon observations of predetermined situations. Comparative findings in those observational studies may lead to impressions, but only after confirmation in study after study do they generally establish causal relationships, and then, only if the criteria discussed below are met.

The observational analytical studies customarily needed to answer environmental and toxicological questions suffer from a number of limiting impediments. Most importantly, they frequently study events that have already occurred, that is, a group of people who have cancer. When they do, variables cannot be controlled prospectively and controls must be developed post facto.

Case-Control Studies

Case-control studies focus upon a population with a particular disorder (cases) and try to assess associated factors (potential causes). Assume, for example, we wanted to know whether coffee consumption caused stomach cancer. We might identify 100 or so individuals with stomach cancer, in whom we are able to get information about coffee consumption. We would then select a matched control group with no cancer—similar (we hope) in all other respects to the study group. To ensure that similarity, we would do careful matching of all variables considered important, as discussed in Chapter 3. We would then determine the extent of coffee consumption in each of the groups and apply proper mathematical analysis to the data to see if there were a difference.

If the group with the cancer had a statistically greater likelihood of comprising coffee drinkers, we would say that we have found an association, in this study, under these conditions, between stomach cancer and the consumption of coffee. We would not have proven a causal connection with this single study for a variety of reasons. First, the connection may have been through a covariate—another factor that was associated with both coffee drinking and the cancer. Second, our controls may have been imperfect. Third, we may have simply identified a chance association, merely a statistical phenomenon, rather than a biologically meaningful connection. Notwithstanding these limitations, the results of these case-control studies commonly appear prominently in our newspapers because their findings, if indicative of a causal relationship, have immediate and significant public health implications. Vitamin E and heart disease, vitamin C and the common cold, coffee consumption and pancreatic cancer, estrogens and breast cancer or heart disease are but a few recent examples.

Case-control studies are retrospective in nature. This means that researchers select a population (the group with disease) at a particular point in time and look backward, or retrospectively, to find a connection with earlier exposures. That fact alone diminishes the quality of controls. After all, how precisely can we probe the details of a person's dietary life, occupational exposure, or smoking habits of 30 years earlier?

Cohort Studies

Cohort studies, another form of analytical observational study, differ from case-control studies in that they study an exposed group, rather than a diseased group. They then compare outcomes with those in an unexposed control population.

Cohort studies may be retrospective or prospective. An example of a retrospective cohort study might be the following. A series of studies were performed to determine whether people who were exposed to benzene in the tire manufacturing industry had a higher incidence of cancers than those unexposed. These studies assessed exposure levels by measuring benzene at various sites in the workplace. They then matched these levels against job descriptions of former workers to select workers with varying lengths and extents of exposure. Because these workers had already been exposed for 30 to 40 years preceding the study, these studies were retrospective. They were cohort studies, because they compared a cohort of exposed individuals with an unexposed control group. A number of such studies found statistically increased incidences of certain leukemias in individuals with higher benzene exposures than in the controls. It was through this study design that this relationship between benzene and leukemia was established.

A prospective cohort study examines a group from the beginning of its exposure. In industry, for example, there may be ongoing studies of individuals with specific chemical exposures compared with a control group without such exposures. Such prospective cohort studies are designed to find out what, if anything, happens to the exposed group that does not happen, or happens less often, to a control group. One of our most important prospective cohort studies examined (and continues to do so) various risk factors in heart disease. The Framingham heart study, begun in the 1950s, has followed nearly 5,200 residents of Framingham, Massachusetts, to assess various potential contributors to heart disease. Numerous factors, including diet, smoking, blood pressure, and weight have been followed and correlated with various heart disease outcomes. It was largely through this study and others like it that risk factors such as diet, smoking, obesity, hypertension, and diabetes have been linked, quite clearly, to heart disease.[4]

Whether the investigator uses a case-control or a cohort study may depend upon the matters being investigated, the nature of available data, financial resources, and other factors. Sometimes a question may be researched equally well with either type of study. For example, studies of birth control pills and thromboembolic vascular disease have been studied both ways. Individuals with the disease being investigated have been matched against controls and compared for birth control pill use (case-control studies). Women taking the pill have been followed and compared with women not taking the pill to compare rates of thromboembolic disease in both groups (cohort studies).

Sometimes the conditions or circumstances require a particular mode of investigation. Rare diseases are best studied by case-control methodologies. If one wants to understand why a group of people contracted a particular disease, a case-control study is needed. If, by contrast, one wants to know whether a particular occupation is associated with a certain excess risk, a cohort study is necessary. Some of the advantages and disadvantages of each of these types of studies is shown in Table 5.2.

Table 5.2
Strengths and Weaknesses of Case-Control and Cohort Studies

Case Control Studies	Cohort Studies
• Advantages	• Advantages
1. Best for rare and long latency diseases.	1. Provides complete data on cases and stages.
2. Relatively quick.	2. Allows study of more than one effect of exposure.
3. Relatively quick.	3. Can calculate and compare rates in exposed and unexposed individuals.
4. Requires relatively few study subjects.	4. Choice of factors available for study.
5. Can often use existing records.	5. Quality control of data.
6. Can study any possible cause of disease.	

Table 5.2 (Continued)
Strengths and Weaknesses of Case-Control and Cohort Studies

Case-Control Studies	Cohort Studies
• Disadvantages	• Disadvantages
1. Relies on recall of existing records about past exposures.	1. Need to study large numbers.
2. Difficult or impossible to validate data.	2. May take many years.
3. Control of extraneous factors incomplete.	3. Circumstances may change during study.
4. Difficult to select suitable comparison group.	4. Control of extraneous factors may be incomplete.
5. Cannot calculate rates.	5. Expensive.
6. Cannot study mechanisms of disease.	6. Rarely possible to study mechanisms of disease.

Adapted from *Maxcy-Rosenau-Last Public Health & Preventive Medicine,* Last, J.M. and Wallace, R.B., Eds., Appleton & Lange, Norwalk, CT, 1992, p.25.

EPIDEMIOLOGY AND DISEASE CAUSATION

Not infrequently, relationships identified in observational analytical studies (either case-control or cohort) are picked up by the popular press and reported upon as though they have proven a causal connection. With next year's contradictory study comes a diminution in the public mind of the credibility of science: "How can they keep changing their minds?" Actually, it is a misinterpretation of the significance of such studies that leads to this misunderstanding. A single observational study, even a very good one, rarely establishes, conclusively, a causal relationship.

Epidemiologists have attempted to clarify the relevance of such studies. Realizing that a single study, or even a few such studies, rarely proves causation, we might ask, "Well, what does?" Hill and others have attempted to answer this. In the mid-1960s as the Surgeon General was immersed in the issue of smoking and cancer causation, a set of criteria was proposed by which the likelihood of a causal relationship (not a mere association) could be assessed. These have come to be known as the Hill criteria after Sir Bradford Hill:

1. Chronological relationship. The exposure must precede the disease if it is to be considered causal.

2. Strength of the association. The greater the magnitude of the association-relative risk of study cohort versus the control group-the more likely the significance.

3. Intensity or duration of exposure. If those with the greatest intensity and duration of exposure have the greater chance of developing the disease, this supports causation. (This is the dose-response effect discussed in Chapter 4.)

4. Specificity of association. If the effect is a specific and unusual one, associated most commonly with the particular potential cause, then this supports causation.

5. Consistency. If many observers, in many experiments, find the same association, this supports causation.

6. Coherence and biological plausibility. The connection between the potential cause and the possible effect must make biological sense. It should also be found in other situations, that is, in experimental animals.[5-6]

The more of these criteria that are met, the more likely it is that an observed outcome is causally connected to the studied potential cause. One can see from these criteria that proving a causal connection between an exposure and a disease is no easy matter and is rarely the result of a single observational epidemiological study. Dr. Angell, in an editorial in the *New England Journal of Medicine*, noting the recent increase in epidemiological research, cautioned against an overinterpretation of the results. The number of studies has increased, the author explained, because many acute diseases have now been dealt with; as a result, we are now facing chronic diseases with multifactorial etiologies—diet, lifestyle, environment, genetics, chronic infections, and other unknown causes. These potential causes can only be studied through epidemiological research. The author warned, however, how difficult the studies can be to interpret and how misleading the findings can be. She noted,

> There is no question that epidemiological studies of risk factors for disease are of growing interest and importance, both for individuals and for the public health. It is important, however, to remember the pitfalls in interpreting them and to be cautious in advising

patients on the basis of single or conflicting studies. This is particularly true of studies purporting to show only weak associations between exposures and disease. These should be evaluated more critically, by researchers and clinicians alike.[7]

THE USES AND MISUSES OF EPIDEMIOLOGY IN ENVIRONMENTAL TOXICOLOGY

Dr. Angell's warning about the interpretation of epidemiological studies is particularly relevant in the arena of environmental toxicology. Here, popular perceptions and public pressures have forced scientific investigations beyond the limits of scientific capabilities. The effects of small amounts of chemicals near waste sites, in indoor air and in drinking water are not easily studied; yet, the public wants answers. These societal concerns are placing greater and greater demands upon the scientific community, asking questions for which there are no answers, demanding answers to questions that defy scientifically valid studies.

State Departments of Public Health are in the middle of these issues and are central to this interface between science and policy. Homeowners want to know whether a variety of physical complaints came about because of their proximity to a nearby waste treatment facility. Their complaints are numerous, vague, widely varied, and subject to enormous reporting bias because of the pervasive perception that the waste site is causing a problem. The Public Health Department, both a political creature and a scientific investigative organization, is caught in a dilemma. Politically, it is required to answer a question that may be scientifically impossible to investigate. One answer may be politically acceptable, but scientifically inaccurate. The other, scientifically correct, may be politically dangerous. This tension between popular demand and scientific accuracy strains the public confidence in science and tests the power of scientists to maintain their integrity and to explain the limits of science to an impatient and sometimes hostile public.

Waste Site Epidemiology

The Pennsylvania Department of Environmental Resources and the U.S. Agency for Toxic Substances and Disease Registry - U.S. Public Health Service (ATSDR) recently studied a group of citizens in central Pennsylvania who complained of a variety of symptoms that they believed were caused by a nearby waste treatment facility.[8] By the time the study took place, the residents had formed an active citizens group and had filed lawsuits. Their story appeared in *Family Circle Magazine* under the title

"Cluster Diseases: Is Your Family at Risk?"[9] The epidemiological study consisted of a symptom survey carried out in the study population of residents living near the waste treatment facility and a nearby control population. This cross-sectional survey asked the participants about symptoms and also asked whether they had ever had cancer. It is of importance, and to their credit, that the researchers went to the actual medical records to check the validity of the answer to the cancer question. The findings were quite revealing. The symptom questionnaire revealed more symptoms in the study group than in the control group. This finding is typical in communities immersed in real or perceived chemical threats. It has no causal meaning, for symptoms would be expected to follow both perceived and real chemical poisoning. The study group also had a 30% higher reported incidence of cancer. However, when medical records were actually reviewed, their medically proven cancer rates were no different from the control group.

Thus, the study group overreported cancer by 30%. Whether they truly believed that they had cancers when they did not, or whether this represented intentional overreporting is not known. It does illustrate, however, the unreliability of questionnaires that ask for the self-reporting of ailments. If even cancer is misreported, how accurate can the reporting of symptoms such as fatigue, headaches, itchy eyes, and insomnia be?

Self-reporting of symptoms is merely one of the experimental design impediments in the study of the health effects of waste sites. This dependent variable—the effect or outcome—is unreliable. Frequently, equally unreliable is the independent variable (potential cause) or the characterization of the actual exposure.

Dr. Philip Landrigan discussed the critical elements of experimental design in waste site epidemiological studies. He noted,

> Evaluation of disease in populations exposed to hazardous waste dumps requires: documentation of the chemicals in a dump; assessment of the materials released from the dump into environmental media; tracing of the probable routes of human exposure (groundwater, air, direct contact or occupational); development, when possible, of individual exposure estimates and/or direct biological assessment of absorption; precise definition of the sub-populations at highest risk of exposure; and the employment of specific and sensitive health outcome indicators....[10]

Here Dr. Landrigan points out the critical elements of experimental design that we discussed in Chapter 3. The independent variable (potential cause or, in this case, exposure) must be known. The dependent variable (outcome) must be clear. Biases must be minimized. In many waste site studies, none of these requirements is met. Exposure is opined, but unknown. Even the route of access from the site to the people is speculative. Endpoints are vague—symptoms, not illnesses.

Most waste site epidemiological studies that have looked for specific and objective measurements of disease have found no relationship between residential proximity to waste sites and the occurrence of serious diseases.[11-14] By contrast, health questionnaires almost always produce more complaints in individuals living near the waste sites, just as occurred in the Pennsylvania study above. The biases inherent in these questionnaires and the various factors that influence reported symptoms have been discussed by many authors.[15-18]

In another health survey in New Jersey, the relationship between perceived (not actual) exposure and the development of symptoms was conclusively demonstrated. When questionnaires were administered to "unexposed" and "exposed" populations in Somerset County, New Jersey, the symptom response of the group that believed it was exposed was substantially higher than that of the control group and exactly the same as that of another "exposed" group in another part of New Jersey. It was later found that the "exposed" Somerset County group was actually not exposed. Their water was not contaminated, but they had believed it to be. Clearly, in that instance, symptom reporting was directly related to perceived exposure and not to actual exposure.[19] Suffice it to say that such questionnaires, now the hallmark of waste site epidemiological surveys, are quite unreliable.

A waste site epidemiological study published in the *American Journal of Industrial Medicine* illustrates every major design flaw of such studies.[20] It also illustrates an insidious danger: despite the fact that it has no scientific relevance, the fact of its publication conveys to the public the false message that simply living near a waste site poses a health threat. This cross-sectional survey attempted to correlate symptoms in a study population with proximity to a former waste site.

The first problem arose in defining the study (exposed) group and the control (unexposed) group. The waste site was found to have at its rim minute levels (10 parts per billion [ppb] or less) of a variety of volatile organic chemicals (VOCs), including benzene, toluene, trichloroethylene, and tetrachloroethylene. These levels, we now know, are only slightly higher than those commonly found in indoor air. Therefore, one cannot truly distinguish between a control and an exposed population. At these levels, everyone is exposed. In an attempt to define a study (exposed) and a control (unexposed) group, the researchers drew concentric circles around the site. Those in the inner circle, closest to the site, were considered exposed. Those in the outer circle, farthest from the site, were considered controls. There were no actual measurements of chemical levels at either of these sites and, judging from the levels at the site itself, neither group's exposure could actually have been influenced by this site. Moreover, even if there had been enough materials to provide an exposure, for inner and outer concentric circles to represent exposure levels would assume no wind. Prevailing winds would actually have moved the VOCs in a noncircular fashion making some of the "controls" exposed and some "exposed" controls. Failing to measure or even to carry out air dispersion modeling rendered this separation of the control and exposed populations inaccurate and misleading.

Next, the researchers noted that during their data analysis they found that a plant with air emissions of dimethylformamide, xylene, and toluene existed in the midst of a "control" area. Because many of the symptoms, eye and nose irritation, for example, in the questionnaire could be influenced by this, they arbitrarily cut out and discarded a section of their control group. It was just as likely, of course, that other sections of either the control or the exposed group were affected by this plant. Since no actual measurements were performed, neither the area nor the individuals affected by the second plant were properly identified.

Finally, the reporting biases inherent in these symptom questionnaires were profound. They were particularly magnified by the fact that the "exposed" group near the waste site belonged to an active citizens organization, and many of whom were claimants in a lawsuit.

Studies such as these are so misleading that they undermine rather than help elucidate our understanding of the science of health effects of contaminants from waste sites. In a very real sense, they play to the perceptions and biases of the public, seemingly offering the imprimatur of science to those perceptions.

"Sick Buildings"

The "sick building" syndrome will be discussed fully in Chapter 13. That certain illnesses and symptoms can be caused by indoor air is irrefutable. But the popular perception of the sheer magnitude and prevalence of "sick buildings" rests in these very difficult areas of epidemiological research. Here, perception frequently plays a greater role than does solid science.

Defining a building as sick often depends upon questionnaires with all of the biases and confounders already described. The measure of the independent variable, such as, factor(s) thought to be responsible for the specific complaints, is frequently inexact, or its relationship to those complaints cannot be accurately known.

Studies in the literature are an admixture of HVAC (heating, ventilation, and air conditioning) data performed by engineers, cross-sectional surveys performed by public health or occupational medicine experts, and industrial hygiene analyses of indoor air constituents. This has produced some fragmentation, each group of separate experts assuming that certain data from the other's field have been clearly proven, when it has not been.

An article written by a human resources and economics expert may project the economic effects of "bad indoor air" and the engineering costs required to reverse that loss. That opinion may be based upon a questionnaire in which individuals are asked to grade the quality of their indoor air. That the worker's perception about the quality of indoor air is an accurate reflection of the reality is assumed by the researchers, and that assumption (which may be quite inaccurate) is used to calculate the economic effects of indoor air upon worker productivity.[21]

Thus, studies of "sick buildings" suffer from fundamental epidemiological difficulties. Studies defining the confounders and biases, assessing the outcomes (such as, symptoms or productivity), and examining potential causes are but a few.

Electric and Magnetic Fields

Another difficult-to-study environmental issue that has been addressed through epidemiological investigation is how electric and magnetic fields affect people. The question posed is, Do high-power lines that produce electric and magnetic fields cause adverse health effects in people? When Wertheimer, in 1979, described a correlation between childhood leukemia and proximity to power lines, public concern began.[22] In fact, this was merely a cross-sectional survey and the independent variable—field strength—had never been measured: it was assumed, based upon wiring configurations. Since then, many other studies have been performed; some finding weak associations between fields and outcomes, such as childhood leukemias, others finding none.[23-29]

In these studies, outcomes can be objectified—cancer is a clear end point, for example. The independent variable, or field strength (either magnetic or electric), is far harder to fix, however. It varies widely from time to time and place to place, even in the same room of the same house. Other variables such as confounders are impossible to control fully. Studies have not shown a dose-response relationship between field intensities and adverse outcomes. No animal model connecting such fields to cancer exists. The biology of this phenomenon is unclear and is not explained by any current understanding of how cancers are produced. Thus, to date, Hill's causation criteria are not satisfied. Electric and magnetic fields have not been shown to cause cancer. The perception of the public is otherwise. Books such as *Currents of Death* and *Cross Currents* have overinterpreted these preliminary observations, averring that the causal relationship is well established, and, moreover, that our government has plotted to withhold the truth from the American public.[30,31]

Occupational Cancers

In the mid-1700s, Sir Percival Pott performed an epidemiological study in which he described the first chemically induced cancer. He had noticed that a rather rare cancer—cancer of the scrotum—seemed particularly prevalent in chimney sweeps. Noting that chimney sweeps removed their clothing before plying their trade, Dr. Pott found that their cancers occurred in an area that came in direct contact with chimney soot and remained unwashed. While Sir Percival could not have known either which chemical or the mechanism by which it caused the cancer, he did correctly determine that something in the soot of chimneys was the cause. Thus began the application of epidemiology to occupational cancers.

Epidemiology has played an invaluable role in uncovering relationships between occupational exposures and various malignancies. Benzene and leukemia, asbestos and mesothelioma and other lung cancers, vinyl chloride and hemangiosarcomas of the liver, bis(chloromethyl)ether and oat cell carcinoma of the lung, hexavalent chromium and lung cancer, radon and lung cancers are but a few of the relationships. Some cohort and case-control studies have investigated relationships between specific occupations and various cancers. Others have examined relationships with specific chemicals. To date, approximately 20 specific chemicals have been linked clearly through strong epidemiological evidence to human cancers. Approximately a dozen specific industries have been clearly linked to excess cancer risks in workers in those industries.

Before these connections were made clearly and convincingly, Hill's criteria, noted above, were satisfied. Numerous studies, not just a number of studies, showed statistically strong associations. They did so consistently. Many times, but not invariably, the increased risk was linked to increased exposures.

Cancer studies, if properly done, have well-defined outcomes (dependent variables): the cancer itself. The independent variable (potential cause), the exposure, is often more difficult to know precisely. Often it can be estimated and, in the case of heavy occupational exposures, can be separated quite distinctly from an unexposed control group. Other influential factors or confounders are impossible to control entirely, but that is so in all epidemiological research.

Typically, associations between chemical agents and malignancies are most readily uncovered when the form of cancer is rare. DES and clear cell adenocarcinoma of the vagina, asbestos and mesothelioma, vinyl chloride and hemangiosarcoma of the liver are examples. Because those cancers are so unusual, even a small increase stands out boldly from the background rate, enabling a clear statistical link to a specific causal agent. Associations between chemical agents and cancers also become apparent when the excess risks are high, even if the cancer is a common one. Smoking and bronchogenic cancer of the lung, asbestos plus smoking and lung cancer are such strong risk factors that their contributions to this common disease also stand out from background rates. By contrast, when cancers are common and risk factors small, relationships can be impossible to find or to confirm. A small relative risk is indistinguishable from no risk at all. The epidemiological literature is overflowing with thousands of articles in which such low-risk relationships between chemical exposures and various cancers have been reported. Whether these reflect actual causation or a statistical curiosity may never be known. What is clear, however, is that such relationships are commonly publicized prematurely, raising fears and concerns with no scientific basis.

SUMMARY

Table 5.3 identifies and rates qualitatively the experimental inaccuracies that plague these areas of research. The quality of the studies with regard to the control of dependent variables, independent variables, and biases is qualitatively rated: usually, sometimes, and rarely. The more often rarely is noted, the less reliable the studies.

Table 5.3
Experimental Inaccuracies

	Independent Variable	Dependent Variable	Biases	Confounders
Waste Site Effects	rarely	rarely	rarely	rarely
Low-Level Water Contamination	rarely	sometimes	rarely	rarely
Electric and Magnetic Fields	sometimes	sometimes	sometimes	rarely
Sick Buildings	rarely	rarely	rarely	rarely
Occupational Cancer	usually	usually	sometimes	sometimes

Community health effects associated with waste sites and sick building studies head the list for unreliability. Environmental epidemiologists are searching intensely for both outcome measures: tools that will enable subtle and specific objective assessments of chemical effects, as well as exposure end points, tools that will permit an objective assessment of, rather than a guess about, a subject's actual exposure. Studies of enzyme systems, of immunological parameters, of neurologic and neurophysiological function are all part of this research landscape. The critical goal is to find more effective, specific, objective, and reproducible ways of measuring independent (exposure) and dependent (outcomes) variables. Until that is accomplished, environmental epidemiology will be plagued by lack of reproducibility, biases, and enormous uncertainties.

REFERENCES

1. McGrew, R.E., *Encyclopedia of Medical History*, McGraw-Hill, New York, 1985.

2. Last, J.M., Ed., *A Dictionary of Epidemiology*, Oxford Press, New York, 1983.

3. Morgenstern, H., Uses of ecologic analysis in epidemiologic research, *Am. J. Pub. Health*, 72, 12, 1982.

4. Dawber, T.R., *The Framingham Study: The Epidemiology of Atherosclerotic Disease*, Harvard University Press, Cambridge, MA, 1980.

5. Hill, A.B., The environment and disease: association or causation?, *Proc. R. Soc. Med.*, 58, 295, 1965.

6. U.S. Department of Health, Education and Welfare, *Smoking and Health: A Report of the Surgeon General*, U.S. Government Printing Office, Washington, DC, 1964.

7. Angell, M., The interpretation of epidemiologic studies, *N. Engl. J. Med.*, 323, 823, 1990.

8. Agency for Toxic Substances and Disease Registry, Division of Health Studies, *Study of Disease and Symptom Prevalence in Residents of Yokon and Cokeburg, Pennsylvania*, ATSDR, Atlanta, 1990.

9. Hales, D., "Cluster diseases: is your family at risk?," *Family Circle Magazine*, April 24, 1990.

10. Landrigan, P.J., Epidemiologic approaches to persons with exposures to waste chemicals, *Environ. Health Perspect.*, 18, 93, 1983.

11. Dunne, M.P., Burnett, P., Lawton, J. and Raphael, B., The health effects of chemical waste in an urban community, *Med. J. Aust.*, 152, 392, 1990.

12. Baker, D.B., Greenland, S., Mendlein, J. and Harmon, P., A health survey of two communities near the Stringfellow waste disposal site, *Arch. Environ. Health*, 43, 325, 1988.

13. Hertzman, C., Hayes, M., Singer J. and Highland, J., Upper Ottowa Street landfill site health study, *Environ. Health Perspect.*, 75, 173, 1987.

14. Janerich, D.T., Burnett, W.S., Feck, G., Hoff, M., Nasca, P., Polednak, A.P., Greenwald, P. and Vianna, N., Cancer incidence in the Love Canal area, *Science*, 212, 1404, 1981.

15. Roht, L.R., Vernon, S.W., Weir, F.W., Pier, S.M., Sullivan, P. and Reed, L.J., Community exposure to hazardous waste disposal sites: assessing reporter bias, *Am. J. Epidemiol.*, 122, 418, 1985.

16. Abramson, J.G., The Cornell Medical Index as an epidemiologic tool, *Am. J. Pub. Health*, 56, 287, 1966.

17. Bond, S.S., Beringer, G.B., Kundin, W.D., et al., Epidemiologic Problems Related to Medical Coverage of New Diseases, presented at annual meeting of the American Public Health Association, Dallas, November 1983.

18. Hopwood, D.G. and Guidotti, T.L., Recall bias in exposed subjects following a toxic exposure incident, *Arch. Environ. Health*, 43, 234, 1988.

19. New Jersey Department of Health, Environmental Health Hazard Evaluation Program (prepared in cooperation with Atlantic County Health Department), *Health Survey of the Population Living Near the Price Landfill, Egg Harbor Township, Atlantic County,* New Jersey Department of Health, New Jersey, 1983, 29.

20. Ozonoff, D., Cotten, M.E., Cupples, A., Heeren, T., Schatzkin, A., Mangione, T., Dresner, M. and Colton, T., Health problems reported by residents of a neighborhood contaminated by a hazardous waste facility, *Am. J. Ind. Med.*, 11, 581, 1987.

21. Woods, J.E., Cost avoidance and productivity in owning and operating buildings, *Occup. Med.*, 4, 575, 1989.

22. Wertheimer, N. and Leeper, E., Electrical wiring configurations and childhood cancer, *Am. J. Epidemiol.*, 109, 273, 1979.

23. Savitz, D.A., Power lines and cancer risk, *JAMA*, 265, 1458, 1991.

24. Savitz, D.A., Case-control study of childhood cancer and exposure to 60-Hz magnetic fields, *Am. J. Epidemiol.*, 128, 21, 1988.

25. Tomenius, L., 50-Hz electromagnetic environment and the incidence of childhood tumors in Stockholm County, *Bioelectromagnetics*, 7, 191, 1986.

26. McDowell, M.E., Mortality of persons resident in the vicinity of electricity transmission facilities, *Br. J. Cancer*, 53, 271, 1986.

27. Fulton, J.P., Cobb, S., Preble, L., Leone, L. and Forman, E., Electrical wiring configurations and childhood leukemia in Rhode Island, *Am. J. Epidemiol.*, 111, 292, 1980.

28. Coleman M.P., Bell, C.M.J., Taylor, H.L. and Primic-Zakelj, M., Leukaemia and residence near electricity transmission equipment: a case-control study, *Br. J. Cancer*, 60, 793, 1989.

29. Michaelson, S.M., Influence of power frequency electric and magnetic fields on human health, *Ann. N.Y. Acad. Sci.*, 502, 55, 1987.

30. Brodeur, P., *Currents of Death: Power Lines, Computer Terminals and the Attempt to Cover Up Their Threat to Your Health*, Simon and Schuster, New York, 1989.

31. Becker, R.O., *Cross Currents*, Tarcher Inc., Los Angeles, 1990.

GENERAL REFERENCES

Andelman, J.B. and Underhill, D.W., *Health Effects of Hazardous Waste Sites*, Lewis Publishers, Chelsea, MI, 1987.

Grisham, J.W. Ed., *Health Aspects of the Disposal of Waste Chemicals*, Pergamon Press, New York, 1986.

Hennekens, C.H. and Buring, J.E., *Epidemiology in Medicine*, Mayrent, S.L., Ed., Little Brown and Co., Boston, 1987.

Last, J.M. and Wallace, R.B., Eds., *Maxcy-Rosenau-Last Public Health & Preventive Medicine,* Appleton & Lange, Norwalk, CT, 1992, p. 25.

Monson, R., *Occupational Epidemiology*, 2nd edition, CRC Press, Inc., Boca Raton, FL, 1990.

6
EARLY DAYS OF ENVIRONMENTAL CONCERNS

DuPont's advertising slogan, "Better things for better living through chemistry," instilled in early TV audiences a sense of pride and reassurance, reactions that today, only 30 years later, seem profoundly inappropriate. Pride was felt, because of fierce patriotism that suffused the post-war American life of the 1950s and early 1960s. Reassurance was felt, because we believed that technology held all of the answers for a new and better life. Technology would conquer diseases and feed the world. In short, technology, of which chemistry was an integral part, would ensure the health and welfare of humanity.

Today, chemicals engender quite a different response: protest, hostility, and fear that they are an inescapable and deadly component of the water we drink, the food we eat, and the air we breathe.

Many factors contributed to the profound and meteoric turnaround in public opinion regarding technology and its chemical by-products. One was a feeling that the promise of technology had fallen short of expectations. Disease, hunger, and poverty still existed. Even worse, the public began to feel betrayed by corporate America and its technologies; instead of fulfilling their promise of social benefits, corporations poisoned the planet they were supposed to help. Thus, in the public mind, the cost/benefit scale tilted away from industries. The products they manufactured, the drugs they developed, the jobs they created and the fertilizers they made to grow foods could not begin to compete with the terrifying images of a dead planet and sick inhabitants. The cost was too great. No backlash in public opinion has ever occurred so quickly or so profoundly.

This drastic shift in the public's view of science and technology was the product of two major social changes: the uncovering and publicizing of serious examples of pollution, and the manipulation of a frightened public by both selfless social reformers and demagogues.

Attention to the threat of chemical poisoning came about gradually and gently at first, and was raised rapidly to a crescendo by a chorus of spokespersons for the public consciousness. One of the earliest such spokespersons was Rachel Carson. Lyrical in its language, her 1960 book, *Silent Spring,* portrayed a vivid, sharp view of a planet in peril. Although short on scientific facts and data, *Silent Spring* raised the American consciousness through its visually acute images. Rachel Carson, a biologist who worked for the Fish and Wildlife Service, focused largely upon plants and animals, less upon threats to humanity. Her sharp descriptions painted a clear and disturbing picture:

> Contamination of our world is not alone a matter of mass spraying. Indeed, for most of us this is of less importance than the innumerable small-scale exposures to which we are subjected day by day, year after year. Like the constant dripping of water that in turn wears away the hardest stone, this birth-to-death contact with dangerous chemicals may in the end prove disastrous.[1]

The environmental movement quickly gained adherents and momentum as others such as Paul Erlich, Barry Commoner, and René Dubos added their voices and their own views to these warnings. The relationship of humanity to the Earth and the Earth to the survival of humanity were themes that spoke to the human spirit. These "holistic" views interrelated and connected all of Earth's creatures and life-forms.

Certain issues raised by these environmental spokespersons were unimpeachable, validated by clear demonstrations of multicolored effluents contaminating rivers in which floated dead fish, and smokestacks spewing forth choking, noxious pollutants. Others were exaggerations and embellishments of facts concerning known biohazards. Many environmentalists stretched the limits of the data and based their arguments on conjecture, opinion, and, sometimes, downright unfounded personal anxieties. Rachel Carson began today's irrational fears of residual pesticides in foods by opining that we were being chronically poisoned, likening this to being victims of the Borgias. Erlich, in his *Population Bomb: Population Control or Race to Oblivion?,* demanded draconian measures to ensure zero population growth, without which, he asserted, the planet and all humanity were doomed.[2] In his 1972 book, *A God Within,* René Dubos assured his readers that Western society was heading toward "racial death" and that technological expansion would "lead to collective suicide."[3]

Shaken by these apocalyptic images, Americans marched to the defense of the environment. But that march broke into a gallop only after the stakes increased. It was the linking of certain chemical toxins to human disease, specifically cancer, that forced the issue to the center of the public and political stage. It was the possibility of public health risks that produced the political will to found the EPA and the Occupational Safety and Health Administration (OSHA).

REFERENCES

1. Carson, R., *Silent Spring*, Fawcett Publishing Co., Greenwich, CT 1962.

2. Erlich, P., *Population Bomb: Population Control or Race to Oblivion?*, Alfred A. Knopf, Inc., New York, 1971.

3. Dubos, R., *A God Within*, Scribner, New York, 1972.

7
CANCER AS THE CATALYST: SCIENTISTS, THE MEDIA, AND ZEALOTRY

Before long, issues involving human health, particularly cancer and its prevention, were paramount. Human health was a strong rallying cry. Although the fate of snail darters (a breed of fish whose existence was endangered by the building of a dam) might coalesce only an activist few, the search for a cure for cancer in our children has far broader appeal. The primary target of the plan, well laid out in best-selling books and in the national media, was the eradication of chemicals in the environment. In his *Politics of Cancer* Epstein claimed that the "recent 'cancer epidemic' has been paralleled by the phenomenal growth of the petrochemical industry."[1] CBS television documentary, *The American Way of Death,* with Dan Rather, suggested that Americans were suffering from a growing cancer epidemic for which chemicals were to blame.[2] Others alleged on national television that "60-90% of all cancers are due to man-made chemicals."[3] Neither allegation was accurate. We were not suffering a new cancer epidemic, nor was there evidence that industrial chemicals were the primary or even a significant cause of cancer. The message, however, was effective. The result was a groundswell of public opinion followed by a political and regulatory frenzy. By the early 1970s, popular demand insisted that we be protected at once from chemical assaults both current and future. Fear of cancer fueled the legislative and regulatory processes.

THE "ENVIRONMENT" AND CANCER

The notion that cancer could be understood in simple terms and might thereby be readily controllable was profoundly enticing (though absurdly simplistic scientifically) and of enormous popular appeal. It came as no surprise when the press became excited over a 1968 International Cancer Conference in Israel in which Dr. John Higginson noted that most cancers appeared to be of "environmental origin."

> While we do not know the etiological factors for many cancers, we are in a position to estimate on theoretical grounds the proportions of all cancers that may be of environmental origin. Calculations would indicate that in the United States approximately 80% of all malignant tumors are likely to be environmentally conditioned and thus theoretically preventable.[4]

What an incredible example of semantic confusion this turned out to be! In the medical and scientific vernacular, "environmental" refers to anything not genetic: stress, diet, and childbearing are examples of environmental influences. Actually, Dr. Higginson and various colleagues at that meeting were discussing the worldwide incidences of different kinds of cancers. They had studied and compared cancer incidences in Singapore, southern China, Ceylon, Jamaica, Iran, Chile, Iceland, Mozambique, Puerto Rico, and elsewhere. They used the term "environmental" to mean all external agents that distinguished each of those societies from the others; most notably, dietary differences. In addition, one of the central conclusions of this conference was that there was little correlation between the extent and nature of industrialization and the incidence or nature of cancers. In many instances the least polluted societies had greater cancer rates, while more polluted ones had lower rates. Man-made chemicals, therefore, actually appeared to play a very minor role, if any. Dr. Higginson and other conferees concluded that diet and cigarette smoking together bore the lion's share of responsibility for most human cancer. It was primarily those two factors that collectively led to that cancer estimate: 80% environmental.

That scientific definition of "environmental" was clouded over by the enthusiasm of popular hope. "Environmental" was misinterpreted to mean "controllable," "man-made" influences. It was a theme that had enormous appeal. Everyone, after all, wanted cancer eliminated. For the relatively new environmental movement, it was a fund-raiser's dream. The broadcast media would have years worth of war stories. Regulatory agencies that did battle with these cancer causers could be ensured of increased governmental funding. Politicians had a marvelous cause; campaigning against cancer offered a winning campaign theme. The body politic in general fell in lockstep squarely behind this popular cause. President Nixon declared a "war on cancer" and made it quite clear to all of us that with sufficient funding the war would be won in the next 10 years. No more cancer! Although many knowledgeable scientists, including Dr. Higginson, recognized this popularization as an overzealous misconception, public sentiment precluded anyone from stepping forward to temper the widespread enthusiasm. Who would listen to a spokesperson against the national consensus that gave direction to this "war on cancer"?

Ten years after that initial conference, when the heat of the frenzy had begun to calm, in an interview Dr. Higginson expressed distress over the mischaracterization of his meaning of "environmental."[5] "Environmental" was not to be equated with "chemical" corrected Higginson.

Finding Carcinogens

The war on chemically induced cancers was launched simultaneously on a variety of fronts: controlling industrial discharges, cleaning up existing contamination, and categorizing and cataloging cancer-causing agents. Researchers sought to unmask new carcinogens—cancer-causing chemicals. In the 1960s only a few agents—most notably radiation and coal tar—were known through epidemiological studies to cause cancers in people. For obvious reasons chemicals had to be tested in animals in order to be classified as potential carcinogens. The laboratories in which animal testing grew at a furious rate were key battlefronts in this war. Testing involved a variety of animals, particularly rodents, often specially bred for genetic homogeneity and susceptibility to chemical carcinogens. Through careful study of these sensitive animals, the search paid off. In 1967 the number of animal carcinogens was 500; by 1976 it was 2,400 and rising. But animal tests were both expensive and slow; thus, new and simpler tests were developed using bacteria and cell cultures. These tests on cellular systems identified the ability of chemicals to cause mutations, thought to be an indicator of carcinogenic potential. The "war on cancer" would begin with a search-and-destroy mission, taking aim at these environmental chemicals.

These tests, while inexpensive and easily done, moved us farther and farther away from clear-cut human relevance. We traded significance to humans for vast quantities of data regarding "suspect" chemicals. This research was able to identify, scientifically, agents capable of producing cell mutations and of causing cancers in experimental animals. What was not clear, however, was just how this information related to man. Animals had to be given large amounts of chemicals (MTD, or maximum tolerated doses), much larger comparatively than what humans would ever ingest, even under extreme conditions. Animals were specially bred to be particularly susceptible to developing cancers. People were not. Often there were sex differences in animal responses. Males might get cancers; females might not. How could that be translated to human beings? Species and strains varied. Rats and mice might be different. BC3BF1 mice reacted differently from Swiss mice. How could these variations be reconciled with projected human responses? Were humans more like female BC3BF1 mice or like male Swiss mice or Sprague Dawley Rats? These issues were perplexing, but the certainty that science was on the right track, that the "war on cancer" could be won if these cancer causers were eradicated propelled the process forward.

The popular call for the elimination of cancer causers required that the limits of science be stretched and that certain assumptions be accepted in the interest of "progress." Animals had to be viewed nondiscriminately as surrogates of humans. Any dosage, no matter how small, had to be considered a threat. The essentiality of these assumptions to the "discovery" process was based upon the uncertainties noted above. It wasn't known whether the occurrence of cancer occurring in heavily dosed, highly sensitive animals meant that humans exposed to lower amounts were at risk, but it had to be assumed that they were. And that concept had to be sold to the public. Early on, the scientifically accurate designator "animal carcinogen" gave way to the term "carcinogen." The distinctions between each species of animal and humans were lost. Any animal in any test became the same as humans under environmental exposure circumstances. These assumptions made intuitive sense, even if they were scientifically flawed. They permitted ready use of available scientific data, and they were "playing it safe." Furthermore, they were impossible to disprove and thereby accomplished the politically effective task of shifting the burden of disproving a human cancer risk onto those who would oppose or soften regulation. A potential carcinogen—based upon animal or even cell culture studies—was now considered guilty until proven innocent.

The Demands of Public Policy: The Limits of Science

The message that cancer was "environmental" and that "environmental" meant "caused by man-made chemicals" was sold with such ferocity to the American public that it became axiomatic. The logical corollary to this axiom was that victory was within ready grasp. Not only was this message wrong and misleading, a setup for disappointment, but once politicized, it led to enormous conflict between political desires and scientific realities. Politicians, through the regulatory agencies, wanted to make good on the promise to their constituents. The regulatory agencies were thus forced to act using data of increasingly marginal scientific relevance, uncertainty, and value to make the regulatory decisions demanded by the political process.

It was particularly this category of cancer-causing chemicals that forced us far beyond the limits of scientific knowledge. In other areas, such as acute toxicity of industrial discharges, it was far easier to reconcile accurately society's requirements with scientific facts. William Ruckleshaus, later commenting upon the subtle scientific demands (particularly concerning potential carcinogens) placed upon the EPA, lamented this when he said, "From its earliest days EPA was often compelled to act under conditions of substantial scientific uncertainty. The full implications of this problem were partially masked at the beginning because the kind of pollution we were trying to control was so blatant...."[6] Major pollution, serious enough to cause acute or clear-cut poisoning—mercury, for example—was easy to find and to correlate with health risk. Subtle, potential cancer causers were not.

The growing incompatibility between the pressures of public policy and the limitations of science has been described by Devra Lee Davis as "basic-science forcing," meaning that the laws and regulations preceded the knowledge base and demanded more than science was prepared to provide.[7] There were many reasons why our basic fund of scientific knowledge became quickly exhausted and unable to contend with the onslaught of public demand and regulatory control. First, of the hundreds of thousands of chemicals in commercial use and their by-products, only a relatively small fraction had ever been subjected to systematic toxicity testing. Thus, for many chemicals, we simply had little or no data. Second, for those chemicals that had been studied, we knew something of acute and chronic toxic effects of relatively high dosages (see Chapter 4). We knew very little about the effects of the smaller amounts more commonly found in environmental settings. Third, when the focus of the regulatory agenda and public perceptions moved away from direct organ toxicity to the risk of cancer, these inquiries operated at the very limits of science. They demanded answers with limited or absent data, even without an understanding of the fundamental principles of cancer. Nonetheless, chemicals were presumed guilty until and unless they had been proven innocent, and scientists were called upon to confirm chemical guilt. As a result, the vast void of data spoke as loudly as if masses of incriminating evidence actually existed.

Impossibility of Proving Safety

In the history of both perception of chemical hazards and of regulatory cancer policy, this issue of presumption of guilt versus the presumption of innocence has played an enormously critical role. Chemical concerns over the past twenty years have been heightened dramatically by the inability of science to prove and ensure absolute safety. Moreover, that inability has served well the political designs of those pressing for ever-increasing protection.

Along the way, both the function and capabilities of science have been grossly misrepresented. Science cannot ever prove categorically that a substance cannot cause cancer. It can only demonstrate that the risk is low. In that respect, science bears no specific guilt or shame, for negatives cannot be proven scientifically or otherwise. Prove, for example, that there is no buried treasure in your backyard. You may dig up the entire yard and declare, "I see no treasure." I'll tell you it's buried deeper. Prove that the Loch Ness monster does not exist. After you have combed the entire lake and not found it, I'll assert that it was hiding. Prove that there are no ghosts, no extraterrestrial visitors to Earth, no dinosaurs still roaming the Earth, that you do not have $1 million stashed away somewhere. It can't be done. These examples illustrate the impossibility of proving the negative. In the case of chemical pollutants, in contrast to those situations, this inability has propelled public policy. The last time that may have happened in America was in 17th century Salem, when those who could not prove they were not witches were burned at the stake.

It is quite understandable that the public's response to the inability of science to disprove a cancer risk has contributed to the pressure for more drastic controls. The axiom that chemicals cause most cancers, coupled with uncertainties about specific chemicals, produce the public's inevitable responses "What we don't know can hurt us," and "Remove all potential offenders."

I have touched upon the role of the media and of scientists involved in public policy in contributing to these popular beliefs and calls for action. The influence of these groups on public opinion about the search for environmental carcinogens warrants further exploration.

Role of the Media

The print media and television both took up the fight against environmental chemicals as the central theme in the war against cancer in the late 1960s and early 1970s. That theme has continued relentlessly to the present day, each new "carcinogen" gaining headlines and space on the evening news. Much of this is overplayed, leaving the public with the false impression that another carcinogen has bit the dust and the total eradication of cancer is near. Love Canal, asbestos, Agent Orange, and Alar are but a few of the recent cancer exposés. The war against these chemicals provides for the media a simple and clear construct for a news piece. They have another attractive characteristic. They represent bad news.

It is clearly true that bad news sells. A plane crash makes headlines; 5,000 uneventful flights do not. Pesticides are news when they threaten children with cancer, not when they perform their intended duty of killing pests. The ravages of Agent Orange made particularly attractive stories, in part because of our collective guilt about our treatment of Vietnam veterans. Asbestos, coal dust and its resulting black lung disease, tampons and toxic shock syndrome, DES and damaged female fetuses were all major stories in their time, identifying health risks associated with corporate America and its products. Some of those stories had elements of truth. Some of those agents did produce injuries and illnesses. When they did, it was often on a smaller scale than the media suggested. In many instances, however, seemingly growing in number in recent years, stories of toxic hazards are exaggerated, released prematurely, or are simply inaccurate. The cranberry crisis of the 1960s; malathion and the spraying of medflies in California; Agent Orange and its effects upon Vietnam veterans; residual pesticides in foods, such as the Alar scare; low-level asbestos such as that found in school buildings; indoor air quality and "sick building" syndrome; and chronic poisonings from low-level environmental contaminants are but a few recent examples. These stories are not all without any scientific foundation. Often, however, the quality of the scientific information is forgotten or intentionally ignored. Positions are put forward that satisfy the goals of special interest and activist groups; controversies and opposing viewpoints are not discussed, and, in some instances, stories with no substantive basis at all are reported.

That would not matter all that much if the media were not so influential in forming and confusing public opinion. Whether or not it reports events and relationships accurately, the media is the consciousness and the source of knowledge for the public at large. Thus, accurate or not, the results of the story are the same: people identify with the story and accept a reported problem as an actual problem. Perceptions develop that may have little or nothing to do with scientific fact.

The extent to which television has become enslaved by its own misinformation was highlighted in a recent piece on *20/20.*[8] To the wonderment of Barbara Walters and Hugh Downs, John Stossel reported a story entitled "Emerging Facts" regarding carcinogens. An extensive interview with Dr. Bruce Ames of the University of California uncovered for the audience the fact that many chemicals naturally occurring in foods cause cancers in experimental animals. This was, in a sense, a reverse exposé in which some of the concerns about man-made carcinogens were put into their proper perspective. The story revealed that the ratio of naturally occurring carcinogens in our diet to those from residual man-made pesticides is 10,000:1. To those of us familiar with the science, this came as no surprise since we had been following this in the scientific literature for many years. But to television audiences generally, as reflected in the comments of the *20/20* hosts, this was a revelation of herculean proportions. It struck at the basic fabric of long-held beliefs, many of which were created, or at least enhanced, by the media itself. In an insightful moment of self-criticism, John Stossel noted, "I think we in the press, many of us, have been irresponsible about these things....We consumer reporters especially often report on this scientist's accusations that this substance causes cancer and make a big scary story out of it without really checking to see how good the research was." He clearly realized that media stories had been one sided, emphasizing hazards, risks, unknowns, and fears while minimizing reassurance, context of exposures, relative risks, and other paranoia-tempering thoughts.

I am not suggesting that the media bear sole responsibility for popular misinformation. The media reflect their sources. Some of those sources have a vested interest in conveying partial truths. In the interest of getting out the story, reporters may not have the luxury of comprehensive research of the type necessary to uncover misinformation. Furthermore, they may have limited expertise; not scientists themselves, they have no real way of distinguishing between good and bad science. Furthermore, in fairness, the media covers a vast range of publications and reporters of enormously variable competence. The accuracy of a story will most likely differ from that of a *New York Times* story.

THE ROLE OF PHYSICIANS AND SCIENTISTS

Less forgivable than the media are the scientists who advise them and who mislead the public. They may do so because they are propelled by a mission, not by science. They may believe that the end justifies the means. Science and medicine commonly move down pathways determined by popular perceptions and concerns. Often those concerns tell scientists what they should be looking for and what they should be doing. The "war on cancer" brought funding for cancer research, thereby encouraging scientific interest. When AIDS sprang into the public consciousness, vast financial resources with concomitant scientific interest shifted to AIDS research. In those instances, public policy, which controls much of the money allocated for research, directed scientific interest through that funding. Not infrequently, the intensity of public interest puts great pressure upon the scientific community. AIDS activists argue that insufficient scientific interest and research have delayed the discovery of a cure. They also argue that bureaucratic impediments prolong the FDA's approval of whatever drugs are being considered for

AIDS treatment. Political pressures can have a significant impact upon the scientific process. They can intensify and direct research into cause and/or treatment. They can accelerate the introduction, perhaps prematurely, of new treatments. Nonetheless, while political pressure can force investigation, it cannot force discovery if science is conducted accurately. The fact that people want and even demand a cure for AIDS will direct research efforts toward finding a cure, but cannot mandate that such a cure be found. The same is true of the environment and its relationship to cancer and other diseases. The quest to explore such relationships can be accelerated through intense funding, but the discovery of answers cannot be mandated. The availability of data and the findings generated by the data, not political demands, are the final arbiters of what is discovered scientifically to be true.

From the early days of environmental activism and health-related chemical regulation, scientists have participated in the process in a variety of ways—some advancing the science, others advancing the cause of popular perception and still others advancing regulatory efforts. Each of these is quite different from the others, though many scientists engaged in these activities would like to see their own motives entirely as truth seeking in the interests of the public. One group of science-educated individuals may contribute to the number of perceived hazards by expressing "well-reasoned," expert opinions, their degrees and presumed expertise granting them authority—the lack of scientific bases for their statements notwithstanding. Another group may have a regulatory role and express regulatory viewpoints in scientific language. The third group often engages in scientific investigation independent of political or policy concerns. The scientific merits of these respective roles are quite clear. The researcher carrying out well-designed investigations is practicing the essence of science. The regulatory scientist who deviates from the data to make regulatory, public policy judgments about what ought to be done is practicing a form of sociological science with some interface in the basic science. That activity is necessary and valuable. However, it must be characterized for what it is. It is not basic science, but rather political science with a dose of scientific underpinning. The public has a clear right to understand this. Finally, the scientist who distorts, exaggerates, or misrepresents in order to advance a political agenda is no scientist at all. Worse, he or she is intellectually dishonest, intentionally misusing credentials to mislead. In the early days of the war on cancer, when "the environment" (aka chemicals) was touted as the primary cause, this group of alarmists with scientific credentials held sway and were heard in books, on television, and in the halls of Congress.

The use of science to advance political goals and agendas is not an unfamiliar theme in the history of science and medicine. Scientists have often used their positions and expertise to advance the party line. So integral is this pattern in the history of our national focus upon cancer and chemicals that a historical review may help put this phenomenon into perspective.

One of the most dramatic instances of politically motivated science, particularly in modern times, took place in Stalinist Soviet Union. Around the time of the Bolshevik revolution, the atom was widely considered to be uniform and predictable, characterized by consistent movement of electrons around a nucleus. Then discoveries in the 1930s and 1940s demonstrated substantial randomness in the movement of electrons, which led Werner Heisenberg, Germany's preeminent nuclear physicist, to formulate his "Principle of Uncertainty," which asserts that there is an irreducible degree of randomness in movements of electrons. This theory and others related to it were echoed by Albert Einstein and Neils Bohr, great Western scientists. While the older Soviet physicists embraced these theories, the younger ones did not, not because of perceived or proven scientific inaccuracies, but because of the seeming incompatibility of this randomness theory with the dialectical materialism of the Marx-Engels-Lenin tradition. Thus, it was politically unacceptable hence, incorrect. Maurice Cornforth, a British communist writer expressed this Soviet attitude in a pamphlet entitled, "Dialectical Materialism and Science":

> So far as bourgeois physical theory is concerned, some of its main difficulties center around the theory of the atomic nucleus.... Bourgeois theory in physics is no more capable of understanding the nature of the atomic nucleus than bourgeois theory in economics was capable of understanding the nature of commodities.[9]

Similarly, M. Mitlin, a member of the Soviet Academy of Sciences (its highest scientific body), attacking the principle of uncertainty, in 1948 described it as a politically motivated doctrine put forth by

> Reactionary flunkies of Anglo-American imperialism in the field of science...and only consistent materialism can purge physics of idealistic tendencies.[10]

It is quite clear that the mainstream Soviet physicists of Stalin's day were hopelessly politicized, so that they were unable and unwilling to examine even scientific principles that were well founded both in theory and experimental observation. Theirs was a science directed by philosophy rather than by data.

During that same period, the Soviet Union spawned its leading biologist, whose genetic theories comported with state doctrine but defied all scientific observations and generally accepted scientific principles. In 1939 Trofim Lysenko, a marginal scientist, but party loyalist, ascended to the top administrative position of the Soviet Academy of Sciences. It was his avowed purpose to reform biology, which was rife with "dangerous Western concepts." Rejecting Mendelian and Darwinian concepts of heredity, genetics and evolution, Lysenko set about to make biological science consistent with communist theory and philosophy. As believed by Mitlin and other physicists, it was the perceived purpose of science to support and bolster the notions and intent of the state. By this point, the world had rejected the Lamarckian notion that genetic change occurred in direct response to environmental demands: that the giraffe's long neck was created by stretching to reach uppermost leaves of trees and that, once achieved, this trait was passed on to offspring. Biological data and theory had, by now, made it quite clear that mutations occurred spontaneously (similar to the randomness theories of physics), and only after such mutations had occurred could they be transmitted to future generations. Soviet economic theory, by contrast, depended upon the buildup of quantitative factors suddenly leading to permanent change. It was that theory with which Lysenko sought to gain consonance as he rewrote genetics. In supporting the agrarian goals of this collective economy, Lysenko's theories argued that cows could be trained to deliver 50 liters of milk per day, and, once trained, they would pass that trait to their offspring. He argued, similarly, that grain yields, once affected by environment, would forever remain in the genetic memory of successive generations. Finally, he argued that wheat would change to rye given appropriate environmental conditions.[11] Claus and Bolander, in their book entitled *Ecological Sanity*, note several key features of Lysenkoism which, I believe, bear some relationship to politically motivated science of today.[12] These include

1. The focus of Lysenkoism was on production. It was necessary to demonstrate the practical relevance of biology to the needs of society.

2. Amassing of evidence was substituted for causal proof as the means of demonstrating the "correctness" of the underlying hypotheses.

3. Ideological zeal took precedence over pure devotion to science; hence those who failed to conform to the tenets of the new biology could be silenced or suppressed as enemies of the truth. Also, it did not really matter if the Lysenko biologists manipulated somewhat their data or their experimental results, since minor falsifications could still support the ideological cause, which represented a higher level of truth than the precise reporting of facts.

Claus and Bolander point out that Lysenkoism, because of its inherent ideological construct, was internally impeccably consistent. Thus, once the belief was adopted, it could be applied broadly and consistently in scientific application. They note that there were common belief systems shared by Lysenkoism and the environmental science movement of the early 1970s that ran contrary to fundamental biological knowledge of Western science. First, they note that from Lysenko, the activist American biologists adopted the notion that the continuous release and presence of chemicals, "in no matter how minute quantities, is bound eventually to result in disastrous changes in living species." Second, that acquired characteristics brought about by exposure to environmental agents will inevitably lead to permanent, inherited biological change. There are, I believe, other more significant relationships between Lysenkoism and the characteristics of the activist environmental scientists.

First of all, the activist environmental scientists have a distinct and undeniable political quality. They desire to further their perceived interest of society in environmental matters and, therefore, bend to provide the imprimatur of science to society's perceptions, scientifically meritorious or not.

Second, and as an outgrowth of the above, the environmental sciences are suffused with certain dogmas, much of it based more in social and political ideology than in scientific knowledge. A recent article by Dr. Bruce Ames described certain of these dogmas.[13] It is instructive to consider them and to compare their characteristics with those expressed by Lysenko as summarized above.

Ames identified elements of this scientifically unfounded but potentially expedient dogma, all of which are essential to the arguments proffered by the scientist cancer activists. These will be discussed more fully in the next chapter. They are that cancer rates are soaring; that only a small number of chemicals are carcinogenic and we can find and eliminate them; that man-made pollutants are present in significant amounts in our environment; that pollution brings cancer and birth defects; and that technology is "doing us in." These dogmas are to the modern scientist cum political activist what Lysenko's dogmas were to him and his colleagues.

Who are the scientists involved with these issues and how do they express their opinions on them? First of all, scientifically our society is far more pluralistic than was the Stalinist Soviet Union. Consequently, there is no single characteristic or quality found in all scientists. Thus, I shall not generalize, but I will make several observations.

The demands of environmental and especially health-related regulation, from their earliest emergence, required of science things that it could not deliver. Protection from future harm or potential hazards fell more into the realm of theory, prognostication, and public policy than of pure science. Science often had no answers; nonetheless, scientists were to deliver them. This meant that scientists had to make judgments that were little more than guess work. Many who did this made it quite clear, and properly so, that they were stepping outside of the scientific role and moving into realms of scientific speculation for policy purposes. Others made this far less clear, blurring the lines between science and policy either intentionally or inadvertently, some because they were devoutly committed to the cause; others because it sold their books. Quite obviously, the stronger the scientific pronouncements, the more weight they carried. Thus, the temptation not to emasculate the message in the interest of scientific integrity was often overwhelming. The ends, many believed, justified the means or, as Claus and Bolanger noted of Lysenkoism, "minor falsifications could still support the ideological cause, which represented a higher level of truth than the precise reporting of facts."[12]

Edith Efron, an extraordinarily harsh critic of this phenomenon and these individuals, was even more pointed in her condemnation:

> The basic scientist, whether he works for the government or a university, is an intellectual explorer in search of truth, and coercion is no part of his repertoire. The "regulatory" scientist, whether he works for the government or at a university, is an intellectual policeman, whose judgments, if accepted by regulators, are backed up by the guns of the state.[14]

Some physician and scientist environmental advocates have moved well beyond the limits of merging science with policy. They use their positions to foment popular and legislative concerns by, distorting facts, and popularizing sensational but scientifically invalid precepts.

The Politics of Cancer, by Dr. Samuel Epstein, rapidly became a best seller in 1978.[1] It was the popular successor to the earlier environmental books of Carson, Commoner, Ehrlich, and Dubos, bringing to the public's mind the notion of cancer as a preventable disease due, overwhelmingly, to excesses of our industrial nation and its waste disposal practices. In its sequel, *Hazardous Waste in America: Our Number One Environmental Crisis*, Dr. Epstein continued this theme. Both books suffered a common defect. While they expressed certain established facts accurately, they both overplayed and exaggerated the extent of scientific knowledge regarding causes of cancer and, particularly, the relationship of cancer to the environment. As a result, both were suffused with ideology exceeding the limits of scientific certainty. Epstein's statement that "cancer is caused mainly by exposure to chemical or physical agents in the environment" was true but misleading. The fact that diet is probably the largest such "environmental agent" was treated minimally in his book. Today's widespread misconception about the extent of naturally occurring carcinogens and mutagens attests to the premature generalizations that characterize this work.

Generalizations are commonly a tool used to incite public opinion. Like racial bigotry, which serves to dehumanize its targets and to unite its opponents, classifying chemicals as "cancer causers" is quite effective in forestalling debate. It glosses over the vast body of scientific literature, ignoring the specifics, keeping the facts from getting in the way, permitting global emotional decisions and beliefs, disregarding completely enormous scientific uncertainty. In many ways it trivializes science and its practitioners; it brushes aside the ongoing and past efforts and conclusions of thousands of dedicated researchers, offering instead simplistic answers and rendering their life's work irrelevant.

Environmental science activists may find themselves in public policy positions in the regulatory agencies. The ideological bent of one of the leading environmental scientists of the 1970s was revealed in Dr. Umberto Saffiotti's discussion of "environmental scientists," about whom he said, "When we review the scientific evidence on a suspected environmental carcinogen, we are in fact, testifying on the circumstances of a suspected mass murder case."[16] With that statement, an ideology is clearly and strongly expressed, but it is hardly one that is likely to lead to scientific objectivity.

Even the most responsible of regulatory scientists are presented with a difficult task in balancing the needs of public policy against the availability of scientific data. Clearly, there is little choice. Decisions have to be made before science is ready to contribute final answers to the process. One can only hope that scientists engaged in this exercise do so responsibly and that they will help nonscientists to understand how the lines are drawn—where science ends and policy begins.

Unfortunately, they commonly do not. The interest of forwarding the goals of public policy demand at least the semblance of expertise. The courts, too, insist upon certain perversions of science. They will strike down regulations that are "arbitrary" or without certain scientific basis. Thus, regulatory scientists are forced to act certain even when they are not, or they may choose to in order to complete their mission.

Of most concerning is that sometimes, it seems, environmental scientists themselves have forgotten the limits of science. Some, so steeped in the now 20-year-old traditions of the health-related regulatory sciences, have lost a clear perspective. Having lived for so long with the dogmas of regulation—"dangers of small quantities," "lack of thresholds," "soaring cancer rates," and the like—they see these as scientific facts, no more aware than the public at large of their roots in social scientific philosophy. I have seen that, as a result of their curricula, graduates of environmental science programs are often unable to distinguish between public policy and science. This is particularly distressing, for if a "scientist" doesn't know what science is, how can a lay person?

REFERENCES

1. Epstein, S.S., *The Politics of Cancer*, Sierra Club Books, San Francisco, 1978.

2. CBS Reports, *The American Way of Death,* October 15, 1975.

3. CBS News, *Face the Nation,* April 18, 1976.

4. Higginson, J., Distribution of different patterns of cancer, *Israel J. Med. Sci.*, 4, 460, 1968.

5. Maugh, T.H., Research news, cancer and environment: Higginson speaks out, *Science*, 205, 1363, 1979.

6. Ruckelshaus, W.D., Risk, science and democracy, *Issues Sci. Technol.*, 1,19, 1988.

7. Davis, D.L., The shotgun wedding of science and law: risk assessment and judicial review, *Columbia J. Environ. Law*, 10, 67, 1985.

8. *Stossel, J., 20/20,* March 18, 1988.

9. Cornforth, M., *Dialectical Materialism and Science*, Lawrence & Weishart, London, 1949.

10. Mitlin, M., *Literaturnaya Gazeta*, November 20, 1948.

11. Lysenko, T.D., Novoe v nauke o biologicheskon vide [A new characterization of the concept of biological sciences, or the fight against the ideological basis of the reactionary Mendel-Morgan theories, the falsification of darwin's teachings], in *Agrobiologia: raboty po vaprosam genetiki, selektsii semenovodstva [Agrobiology, Contributions to the Problems of Genetics, Selection and Breeding for Improved Seed Varieties]*, Lenin Akad. Agr. Nauk, Moscow, 1951, 604.

12. Claus, G. and Bolander, K., *Ecological Sanity*, David McKay Company, New York, 1977.

13. Ames, B.N., What are the major carcinogens in the etiology of human cancer? Environmental pollution, natural carcinogens, and the causes of human cancer: six errors, in *Important Advances in Oncology*, DeVita, V.T., Jr., Hellman, S. and Rosenberg, S.A., Eds., J.B. Lippincott, Philadelphia, 1989.

14. Efron, E., *The Apocalyptics*, Simon and Schuster, New York, 1984.

15. Epstein, S.S., *Hazardous Waste in America: Our Number One Environmental Crisis*, Sierra Club Books, San Francisco, 1983.

16. Saffiotti, U., Scientific basis of environmental carcinogenesis and cancer prevention: developing an interdisciplinary science and facing its ethical implications, *J. Toxicol. Environ. Health*, 2, 1445, 1977.

8
ENVIRONMENTAL CHEMICALS: COMMON BELIEFS VERSUS SCIENTIFIC KNOWLEDGE

It does not take many events like the one in Bohpal, India, on December 12, 1984, for people's fears of chemicals to rise to the surface. Worth more than a thousand words, these television and newspaper images of many hundreds of innocent villagers dead or gasping for breath emblazoned themselves indelibly in the public consciousness. Every publicized toxic event, situation, or substance builds incrementally in the public mind, ensuring an already-suspicious and worried populace that toxic threats are pervasive. The press and the public, limited in scientific understanding, cannot readily discriminate among different kinds of toxicity. There is little understanding of the differences in scope or seriousness of different occurrences. It doesn't much matter whether the event was a poisonous chemical cloud, as in Bohpal, or infinitesimal (and likely harmless) residues of the pesticide Alar in apples. To most people, they are all the same: poisons of industry. Thus, Three Mile Island, Chernobyl, black lung disease, radon, asbestos, lead, toxic waste sites, Love Canal, Times Beach, Agent Orange, and contaminated drinking water all add incrementally to the popular perception of pervasive toxicity.

There is also, I believe, a clear link between environmental concerns generally and perceptions about chemical and other toxic hazards. Thus, publicity about the greenhouse effect, global warming, the reduced ozone layer, and medical waste in the oceans all contribute to the image of an increasingly contaminated and dangerous planet. With this backdrop, this public mind-set, nearly any toxic claim or report is readily and uncritically accepted—another piece of data supporting what we already know about toxic threats.

The answer is not to end discussion and debate about important environmental issues, but to place them into context and to weigh them according to the severity of the threat. The danger when every report is accorded equal status lies in the inability to prioritize. Unless people know what to take most seriously, they cannot decide which actions to take first or where to spend scarce resources. Eventually, we may be so numbed by one environmental Armageddon after another that we throw up our hands, light a cigarette, and give up all together. This effort to put toxic phenomena into proper and accurate perspective begins with an exploration of certain key areas in which popular perceptions and scientific realities diverge. It is impossible to discuss all of these, but a number stand out.

THE RECEDING ZERO

It is commonly believed that any amount of unwanted chemicals in our air, food, or water is harmful. That is not so. The mere presence of chemicals is insufficient to warrant alarm. Rather, their amounts or concentrations are central. People are naturally distressed when they hear that trichloroethylene is in their drinking water—water they believed pure—or that toluene is in their indoor air—air they believed pure. The fact is that the very definition of purity has been turned upside down by our ever-increasing ability to measure smaller and smaller amounts of substances. As that ability has increased, we have come to realize that "purity" is elusive, permanently limited by the intensity of our analytical abilities. This has been called by Dr. Sidney Shindell "the receding zero."[1]

Enhanced analytical capabilities have enabled us to measure minute quantities of substances. Just as we think there is no more present, lo and behold, new technology picks up another trace. Thus, we continually chase "the receding zero." One year "none" is found. The next year "some" is found. It was estimated by Dr. George Koelle that if one pint of water were spilled into the ocean, once it mixed fully and completely, 5,000 molecules of that water would be present in every pint of water taken anywhere in the world. The same analogy holds true for any and every chemical either found in nature or imported by humans. Every time we wash a chemical down the sink, some fraction of either that chemical or its breakdown products will eventually be found in every glass of water in the world. Thus, the fact is that everything mixes with everything else; essentially, nothing is uncontaminated.

The same analogy can be made for airborne substances. We drive our cars, which put out thousands of separate chemicals in each puff of exhaust. We dryclean our clothes—the chemicals evaporate into the air. We run lawn mowers, heat our homes, burn wood in our fireplaces, smoke cigarettes, treat our lawns with chemicals, paint our houses, polish our wood tables, and the list goes on. Each time, molecules enter the air. Depending upon what they are, they may be degraded or changed chemically or they may not. In any case, some fraction of them or their breakdown products remains in the environment to be transported throughout and mixed with the molecules of the world.

This may not be comforting. It is, in fact, the image that sparks broad interest in environmental awareness and controls, and it should. But it is also a fact that puts into perspective the issue of contaminated water and air as we search for the receding zero.

Because of this equilibration between human sources of chemicals and the world about us, no environmental substance—air, dirt, or water—can or ever will be free of chemicals. Reducing levels to none is like proving the negative. It cannot be done. At some level of measurement, perhaps only one molecule, "some" will always be found as long as the material exists on the earth. Continually pressing for the eradication of the last trace is like a dog chasing its tail. It keeps going around and around in circles never reaching its target, never accomplishing his task. To stop chasing our tails, we must ask the critical question, not "Is any present?" but "Is any meaningful amount present?"

To answer that question we must harken first to basic principles of toxicology. We learned in Chapter 4 that "the dose makes the poison." We need be concerned about toxicity when the amount of a chemical begins to approach toxic levels. As long as it is far away from such levels, it is not likely to cause harm. The other aspect of toxicity involves the potential of risk, particularly the risk of cancer. This is discussed more in Chapter 10 when we consider risk assessment, but will also be covered to some extent in the discussion to follow. It is these considerations—the quantities to produce toxicity and meaningful risk—that tell us how relevant small amounts of contaminants are. Remember, that they cannot be eliminated, so their relevance must be understood.

Putting these exposures into context can also be assisted by an understanding of how little we can measure and how small the quantities of commonly reported environmental contaminants actually are. Chemicals found in drinking water and air are commonly measured in ppb or less. One part per billion is difficult to visualize. Analogies sometimes help. Some of those are summarized in Table 8.1.

Table 8.1. Examples of One Part Per Billion

1 part per billion (ppb)
1 ounce in 64,000 pounds
500 feet on a path to the sun
1 hair on 10,000 heads

Sometimes substances are measured in parts per million (ppm) (particularly in the workplace). This is 1,000 times greater than a ppb. Environmental contaminants are often found in even smaller concentrations, parts per trillion (ppt) (1,000 times smaller than a ppb), or even parts per quadrillion (ppq) (1,000,000 times smaller than a ppb).

These very small numbers can be made to sound like a great deal simply by changing the units of measure. One ppt is 1,000 ppq. Even though 1,000 ppq is exceedingly small, all of those zeros make it sound quite large indeed. One can make 1 ppm sound bigger by calling it 1,000 ppb or 1,000,000 ppt or 1,000,000,000 ppq, all proper representations of the same number. Analogously, the same unit changes can be made to portray concentrations of substances in other terms: g/m^3, for example. One g/m^3 equals 1000 mg/m^3 equals 1,000,000 $\mu g/m^3$, and so on. These equivalencies are shown in Table 8.2. We must be careful not to confuse the size of the number with the quantity of the contaminant. The unit, not the numerical size, determines the quantity.

Table 8.2
Equivalent representations of the same concentration illustrates that the size of the number does not represent the amount of material.

1 ppm	=	1,000 ppb	=	1,000,000 ppt	=	1,000,000,000 ppq
1 g/m^3	=	1,000 mg/m^3	=	1,000,000 $\mu g/m^3$	=	1,000,000,000 ng/m^3

PARTS PER BILLION IN OUR BODIES

It is commonly believed that if any amount of a man-made chemical is found in our body, we have been poisoned. That is not so. The receding zero applies not only to chemicals in the external environment, but also to chemicals in our own bodies. We are in equilibrium with our environment. We drink the water, breath the air, and eat the meat, fish, and crops that, in turn, equilibrate with their environments. It stands to reason that if air, water, and soil contain some molecules of all chemicals on Earth, then so, too, would we inside our bodies. In actuality we do. The EPA commissioned a study known as the Broad Scan Study.[2] In this study, fat samples were taken at random from thousands of individuals undergoing surgery or who had died (not of chemical poisoning), and a variety of chemical substances were measured. A vast number of environmentally present chemicals were measured at some level in the majority of samples tested. A summary of some of those findings is presented in Tables 8.3, 8.4, and 8.5.

These tables indicate the average levels of a variety of chemicals found in the fat of Americans. The units of measure are, once again, very small (ng/gm or ppb). Notice, however, if we wanted to make the amount appear larger, we could report the number in numbers of molecules in our bodies. Look at ethylbenzene, for example. The level of ethylbenzene found in the tested Americans ranged from 1 to 280 ng/g. Let's assume only 100 ng/g of ethylbenzene was found. The 100 ng/g of fat or ppb level translates to 600,000,000,000,000 molecules of ethylbenzene. These measurements are merely the chemicals that were part of this study. Had the EPA chosen to look for thousands of others, they would have found them. Again, the question is not "Are they present?" Inevitably they are and always will be at some level. The question is "Are they present at a dangerous level?"

Since the real underlying question is "Does this put us at risk, particularly, of getting cancer?" we can address these questions in a number of ways.

THE CONTEXT OF EXPOSURE: MAN-MADE VERSUS NATURAL CARCINOGENS

It is commonly believed that when it comes to chemicals, particularly where cancer is concerned, natural is good or benign, while man-made is bad. That is not so. Dr. Bruce Ames of the University of California helps us put this issue of chemical carcinogens into proper perspective. He notes that hundreds of tested chemicals, including numerous natural (not man-made) components of fruits, vegetables, and meats, are carcinogenic in experimental animals.[3-5] In fact, 50% of all chemicals tested are carcinogenic in one test or another, whether they are natural or man-made. Dr. Ames further notes that "Americans ingest in their diet at least 10,000 times more natural pesticides (by weight) as man-made pesticide residues; that these "are by far the main source of toxic chemicals ingested by humans"; and because their concentrations are in parts per thousand or more, rather than the usual ppb for synthetic pesticide residues, "our diet is likely to be very high in natural carcinogens."[6]

These natural carcinogens are found in a variety of sources. Common are the natural pesticides, chemicals made by plants to protect themselves against insects. Some that have been found are in mushrooms, parsley,

Table 8.3 Incidence of Detection of Target Volatile Organic Compounds in the NHATS FY82 Composite Specimens

Compound	Frequency of observation (%)	Wet tissue concentration (ng/g)[b]
Chloroform	76	ND (2)[a] - 580
1,1,1—Trichloroethane	38	ND (17) - 830
Bromodichloromethane	0	ND (21)
Benzene	96	ND (4) - 97
Tetrachloroethene	61	ND (3) - 94
Dibromochloromethane	0	ND (1)
1,1,2—Trichloroethane	0	ND (1)
Toluene	91	ND (1) - 250
Chlorobenzene	96	ND (1) - 9
Ethylbenzene	96	ND (2) - 280
Bromoform	0	ND (1)
Sturene	100	8-350
1,1,2,2—Tetrachloroethane	9	ND (1) - 8
1,2—Dichlorobenzene	63	ND (0.1) - 2
1,4—Dichlorobenzene	100	12-500
Xylene	100	18-1,400
Ethylphenol	100	0.4-400

[a] ND = not detected. Value in parentheses is the estimated limit of detection.
[b] The exact isomers were not determined.

Source: U.S. Environmental Protection Agency, Office of Toxic Substances, *Broad Scan Analysis of the FY82 Survey Specimens, Volume 1, Executive Summary*, National Technical Information Service, Springfield, VA, 1986.

Table 8.4 Incidence of Detection of Target Semivolatile Organic Compounds in the National Human Adipose Tissue Survey (NHATS) FY82 Composite Specimens

Compound	Frequency of observation (%)[a]	Range of observed lipid concentration (ng/g)
Dichlorobenzene	9	ND (9)b - 57
Trichlorobenzene	4	ND (9) - 21
Naphthalene	40	ND (9) - 63
Diethyl phthalate	42	ND (10) - 970
Tributyl phosphate	2	ND (44) - 120
Hexachlorobenzene	76	ND (12) - 1,300
β-BHC	87	ND (19) - 570
Phenanthrene	13	ND (9) - 24
Di-n-butyl phthalate	44	ND (10) - 1,700
Heptachlor epoxide	67	ND (10) - 310
trans-Nonachlor	53	ND (18) - 520
p,p[1]-DDE	93	ND (9) - 6,800
Dieldrin	31	ND (44) - 4,100
p,p[1]-DDT	55	ND (9) - 540
Butylbenzyl phthalate	69	ND (9) - 1,700
Triphenyl phosphate	36	ND (18) - 850
Di-n-octyl phthalate	31	ND (9) - 850
Mirex	13	ND (9) - 41
tris(2-Chloroethyl)phosphate	2	ND (35) - 210
Total PCBs	83	ND (15) - 1,700
Trichlorobiphenyl	22	ND (9) - 33
Tetrachlorobiphenyl	53	ND (9) - 93
Pentachlorobiphenyl	73	ND (21) - 270
Hexachlorobiphenyl	73	ND (19) - 450
Heptachlorobiphenyl	53	ND (19) - 390
Octachlorobiphenyl	40	ND (20) - 320
Nonachlorobiphenyl	13	ND (18) - 300
Decachlorobiphenyl	7	ND (22) - 150

[a] Sample size = 46 composites.
[b] ND = not detected. Value in parentheses is the estimated limit of detection.

Source: U.S. Environmental Protection Agency, Office of Toxic Substances, *Broad Scan Analysis of the FY82 Survey Specimens, Volume 1, Executive Summary*, National Technical Information Service, Springfield, VA, 1986.

Table 8.5
Lipid-Adjuster Concentration of polychlorinated dibenzo-p-dioxin (PCDD) and dibenzofuran (PCDF) in the NHATS FY82 Composite Specimens

Compound	Frequency of detection (%)	Mean concentration [a] (pg/g)	Range of detection (pg/g)
2,3,7,8- TCDD	76	6.2 ± 3.3	ND (1.3)[c] - 14
1,2,3,7,8-PeCDD	91	43.5 ± 64.5	ND (1.3) - 5,000
HxCDD[b]	98	86.9 ± 83.8	ND (13) - 620
1,2,3,4,7,8,9-HpCDD	98	102 ± 93.5	ND (26) - 1,300
OCDD	100	694 ± 355	19 - 3,700
2,3,7,8-TCDF	26	15.6 ± 16.5	ND (1.3) - 660
2,3,4,7,8-PeCDF	89	36.1 ± 20.4	ND (1.3) - 90
HxCDF[b]	72	23.5 ± 11.6	ND (3.0) - 60
1,2,3,4,6,7,8-HpCDF	93	20.9 ± 15.0	ND (3.5) - 79
OCDF	39	73.4 ± 134	ND (1.2) - 890

[a] Mean concentration calculated using trace and positive quantifiable values.
[b] Reference compounds not available to specify isomers.
[c] ND = not detected. Value in parentheses is the estimated limit of detection.

Source: U.S. Environmental Protection Agency, Office of Toxic Substances, *Broad Scan Analysis of the FY82 Survey Specimens, Volume 1, Executive Summary*, National Technical Information Service, Springfield, VA, 1986.

basil, parsnips, fennel, pepper, celery, figs, mustard, and citrus oil. These no doubt represent only a small fraction of the total natural carcinogens, because relatively few have been tested to date. Besides natural pesticides, foods may contain other types of carcinogens. Cooked meat, particularly if browned as in a barbecue, is loaded with carcinogenic nitrosamines. Coffee contains about 500 μg of hydrogen peroxide and methylglyoxal (both animal carcinogens); a cola drink contains about 2,000 μg of formaldehyde (again, an animal carcinogen). Peanuts and peanut butter commonly contain aflatoxin, a potentially carcinogenic mold. Table 8.6 lists some of the carcinogens that have been found in a variety of foods.

Matching the low levels of potentially carcinogenic environmental chemicals against these very high levels of dietary carcinogens tends to put some perspective on the matter. Threats from the air, water, and residual man-made chemicals in foods are trivial by comparison to potential threats from natural dietary constituents.

In an interesting and relevant comparison, Dr. Ames compares the carcinogenic propensity of tetrachlorodibenzodioxin (TCDD) to alcohol and broccoli. First, comparing TCDD with alcohol, he notes that alcohol is known to cause both cancer and birth defects in human beings, albeit weakly. TCDD, by contrast, is not a clearly established human teratogen or carcinogen. Then, considering the relatively enormous quantities of alcohol consumed, in a beer or two a day, for example, and comparing that with the millions of times less of TCDD maximally present in our diets, he finds the concern over dietary TCDD overstated. The equivalent carcinogenic potential would equate to the cancer—causing potential of one beer every 5 months. Dr. Ames carries out a similar analysis in comparing TCDD with members of the Brassica vegetable family, which include broccoli, cauliflower, and cabbage. They contain a natural carcinogen known as indole carbinol. This chemical binds to the very same receptors in cells as does TCDD. Stomach acid actually causes these indole carbinols to convert to a group of chemicals structurally quite like dioxins, which clearly mimic the effect of dioxins on cellular receptors and enzyme systems. Dr. Ames has estimated that customary consumption of these vegetables produces this chemical in amounts billions of times greater than the EPA TCDD food limit. Roughly speaking, one portion of broccoli would provide 20 million times the carcinogenic potency of a daily permissible TCDD intake.

Table 8.6 Carcinogens Found Naturally in Foods

Food	Potential Carcinogen
Peanuts and peanut butter	Aflatoxin
Brown mustard	Allyl isothiocyanate
Basil	Estragole
Mushrooms	Hydrazines
Beer	Ethyl alcohol
Bacon	Dimethylnitrosamine
Shrimp	Formaldehyde
Bread	Formaldehyde
Strawberries	Benzene
Coffee	Dicarbonyl aldehyde methylglyoxal
Broccoli	Indole carbinol
Cabbage	Indole carbinol

CANCER RATES

It is commonly believed that cancer rates are soaring and that man-made chemicals are to blame. Neither is true. If man-made chemicals were causing epidemics of cancer, we should see those by now. We do not. While some cancer rates are rising-primarily those due to cigarette—smoking and to natural dietary factors such as fat—others are falling.

Table 8.7 delineates changes in cancer incidences from 1973 to 1987 for common sites. Also given are known and likely causes of those

Table 8.7
CANCER TRENDS AND CAUSES
These are the trends and suspected causes of various cancers.

Type	Annual New Cases	% Change 1973-87 (Incidence)	% Change 1973-87 (Mortality)	* Known or Likely Causes
Lung	157,000	31.5	34.1	Tobacco
Breast	150,900	24.2	2.2	Hormones
Prostrate	106,000	45.90	7.2	Testosterone
Bladder	49,000	12.3	-22.7	Tobacco
Uterus	33,000	-26.1	-19.8	Estrogen
Oral	30,500	-1.3	-16.2	Tobacco, Alcohol
Pancreas	28,100	-5.6	-2.0	Tobacco
Leukemia	27,800	-10.2	-5.6	X-rays
Melanoma	27,600	83.3	29.8	Sunburn
Kidney	24,000	27.0	12.9	Tobacco
Stomach	23,200	-20.5	-29.4	Salt, Tobacco
Hodgkins Disease	7,400	-15.9	-49.5	Unknown
Cervix	13,500	-36.4	-39.6	Papilloma virus

Delineates changes in cancer incidences from 1973 to 1987 for common sites. Also given are known and likely causes of those malignancies.

* Known and likely causes do not represent the principal causes, only factors identified by the International Agency for Research on Cancer as causing at least some cases.

Data taken from *Science Magazine*, November 22, 1991

malignancies. Henderson and colleagues[7], in an excellent review, noted,

> The majority of the causes of cancer (such as tobacco, alcohol, animal fat, obesity, ultraviolet light) are associated with life-style, that is, with personal choices and not with the environment in general. The widespread public perception that environment is a major cancer hazard is incorrect.

Overall, our health and longevity are better than ever in our history and there is no convincing evidence that chemical pollutants have contributed to any increases in cancer.

One reason that the chemical carcinogens and the even greater amounts of natural carcinogens do not produce cancers in all of us lies in our enormous resistance to these diseases. We have a wide range of protective mechanisms that inactivate chemical carcinogens, reverse early changes, and repair damaged DNA.

PUTTING EXPOSURES INTO PERSPECTIVE

To make meaningful decisions about controlling chemical exposures we must be able to focus upon those chemicals and those situations of greatest threat.

To help do this, Dr. Ames has developed extensive rating scales in an attempt to put into an ordered perspective the relative cancer risks posed by a variety of daily exposures. This Human Exposure Rodent Potency (HERP) index takes into account the carcinogenic intensity as measured in experimental animals and the extent of exposure in people. When he does this, Dr. Ames notes that certain industrial exposures such as EDB and formaldehyde get a high rating and deserve serious attention. By contrast, many environmental exposures are vastly (1,000 to 100,000 times) less significant than are daily exposures to carcinogens in foods. Table 8.8 summarizes some of this ranking scheme.

SUMMARY

The popular notion that modern society is hazardous to health is founded upon perceptions rather than scientific reality. The corollary—toxins are the primary or a major contributor to this decline in health—is equally unsupported. Deeply held beliefs of this sort are not changed easily, but as fiscal constraints force precise allocation of environmental resources, a health focus will guide the way. That focus must prioritize effectively, spending money where it will do the most good, not wasting it on perceived but unproven hazards. If the public is to participate in these decisions, these lines between perception and scientific reality must be brought into clearer focus.

The underlying message of risk used here is a HERP value. The measure of rodent potency is the milligrams of substance per kilogram of rodent body weight necessary to produce cancer in one-half the rodents, given daily exposure over the rodents' lifetime. Human exposure is measured by the daily consumption indicated in the table per kilogram of human body weight. In Table 8.8, the HERP values have been normalized with respect to the HERP value for water.

Table 8.8 HERP Ranking for Possible Carcinogenic Hazards

Possible Hazard: HERP %	Source/Daily Human Exposure	Carcinogen
	Water	
0.001	Tap water, 1 liter	Chloroform
0.004	Well water, 1 liter (worst well in Silicon Valley)	Trichloroethylene
	Risks Created by Mother Nature	
0.03	Peanut Butter, 1 sandwich	Aflatoxin
0.01	Mushroom, 1 raw	Hydrazines, etc.
2.8	Beer, 12 ounces	Ethyl alcohol
4.7	Wine, 1 glass	Ethyl alcohol
0.0003	Coffee, 1 cup	Hydrogen peroxide
0.03	Comfrey herbal tea, 1 cup	Symphytine
0.4	Bread, 2 slices	Formaldehyde
2.7	Cola, 1	Formaldehyde
0.09	Shrimp, 100 grams	Formaldehyde
0.003	Cooked bacon, 100 grams	Dimethylnitrosamine, diethylnitrosamine
0.06	Cooked fish or squid, broiled in a gas oven, 54 grams	Dimethylnitrosamine
0.07	Brown mustard, 5 grams	Allyl isothiocyanate
0.1	Basil, 1 gram of dried leaf	Estragole
0.02	All cooked food, average U.S. diet	Heterocyclic amines
0.2	Natural root beer, 12 ounces (now banned)	Safrole
	Food Additives and Pesticides	
0.06	Diet cola, 12 ounces	Saccharin
0.0004	Bread and grain products, average U.S. diet	Ethylene dibromide
0.0005	Other food with pesticides, average U.S. diet	PCBs, DDE-DDT
	Risks Around the Home	
0.6	Breathing air in a conventional home, 14 hours	Formaldehyde, Benzene
2.1	Breathing air in a mobile home, 14 hours	Formaldehyde
0.0008	Swimming pool, 1 hour (for a child)	Chloroform
	Risks at Work	
5.8	Breathing air at work, U.S. average	Formaldehyde
	Commonly Used Drugs	
16	Sleeping pill (Phenobarbital), 60 milligrams	Phenobarbital
0.3	Pain relief pill (Phenacetin), 300 milligrams	Phenacetin

Source: Ames, B.N., Magaw, R.G. and Gold, L.S., Ranking possible carcinogenic hazards, *Science,* 236, 271, 274, 1987.

REFERENCES

1. Shindell, S., "The Receding Zero," paper presented at American Council of Science and Health Conference, November/December, 1985.

2. U.S. Environmental Protection Agency, Office of Toxic Substances, *Broad Scan Analysis of the FY82 Survey Specimens, Volume 1, Executive Summary*, National Technical Information Service, Springfield, VA, 1986.

3. Ames, B.N., Mutagenesis and carcinogenesis: endogenous and exogenous factors, *Environ. Mol. Mutagen.*, 14 (Suppl. 16), 66, 1989.

4. Ames, B.N., Magaw, R. and Gold, L.S., Ranking possible carcinogenic hazards, *Science*, 236, 271, 1987.

5. Ames, B.N., What are the major carcinogens in the etiology of human cancer? Environmental pollution, natural carcinogens, and the causes of human cancer: six errors, in *Important Advances in Oncology*, Devita, V.T., Hellman, S. and Rosenberg, S.A., Eds., J.B. Lippincott, Philadelphia, 1989.

6. Ames, B.N., Dietary carcinogens and anti-carcinogens: oxygen radicals and degenerative diseases, *Science*, 221, 1256, 1983.

7. Henderson, B.E., Ross, R.E. and Pike, M.C., Toward the primary prevention of cancer, *Science*, 234, 1131, 1991.

9
THE RISE OF TOXIN-RELATED REGULATION

INTRODUCTION: THE PURPOSE OF REGULATION

The immutable and central theme of all regulatory activity, both health related and nonhealth related, is protection: protection of viewers from false advertising; protection of consumers from fraudulent contracts; protection of airline passengers from excessive prices. In environmental and occupational regulation much of the focus is centered upon health protection, not just of humans, but of fish, plants, and other wildlife as well. Ultimately, the most compelling and (politically) most successful arguments supporting health-related regulation and legislation address issues of public health, though today other environmental health issues and even the health of laboratory animals—animal rights—have significant momentum.

THE BIRTH OF THE AGENCIES

In the mid-1960s, new environmental concerns galvanized both public opinion and the political process and accelerated the implementation of environmental controls at a truly staggering rate. Before 1970 there were no EPA; no OSHA; no National Institute of Occupational Safety (NIOSH); no National Toxicology Program (NTP); no Resource Conservation and Recovery Act (RCRA); no Comprehensive Environmental Response, Compensation and Liability Act (CERCLA or Superfund); no National Cancer Act (NCA); and no Toxic Substance Control Act (TSCA). Prior to 1970 very few regulations dealt with health risks: the Food, Drug and Cosmetics acts, the Federal Insecticide, Fungicide and Rodenticide Act (FIFRA), the Clean Air and Clean Water Acts (CAA, CWA), and various state public health regulations. Most agencies, regulations, and laws dealing with environmental exposures to potentially toxic substances came about in the past twenty years, most more recently than that. More important, the health risk focus changed entirely during the past two decades. Zimmerman traces these changes (particularly in environmental regulation) from attempts to lessen acute health effects through discharge controls designed to minimize risks, to the more recent focus on cleanup of existing potential hazards.[1] Page characterized the emerging legislation and regulation of the 1970s by noting the changes in risk focus:

- New concerns over long-term, life-threatening diseases (largely cancer) rather than upon more acute or short-duration disorders.
- Concentration upon numerous, rather than just a few chemicals.

- Concern over minute levels of chemicals.
- Concern over latent effects, rather than simply over short-term effects.
- Concern over more subtle effects, rather than only over readily observable effects.[2]

During the early 1970s, regulatory efforts focused quite heavily upon major pollution. There the science was clear. Data supported, quite directly, the political decisions. Beginning in the early 1970s and accelerating through the 1980s, regulatory demands grossly outpaced scientific answers. The toxicological sciences grew markedly, but not nearly as fast as toxicological regulations. The 1990s will likely see some change. Scientific knowledge is beginning to catch up with regulatory demand.

WHEN SCIENCE RAN OUT OF ANSWERS

The 1970s produced great enthusiasm and promise for a clean new world. Before long, however, the public and the regulators confronted the harsh reality that public enthusiasm exceeded both information and the scientific wherewithal to realize America's environmental dreams. Winston Churchill described such phenomena with the expression "from the wonderful cloudland of aspiration to the ugly scaffolding of attempt and achievement." Some of that "ugly scaffolding" was built upon the sheer enormity of the task, some upon the incalculable economic costs to the country, and some upon the vast uncertainties and gaps in scientific understanding that simply could not meet the demands of public policy.

As the Agencies were established, they were charged with the protection of public health. Whether or not information existed that could accomplish that task was not considered important. The task was clear and the charge irrefutable. Each act that pertained to public health contained three major categories that demanded action despite the extent of limitations of knowledge. Some of these statutes are listed in Table 9.1.

Here are delineated the statutory provisions, the level of certainty, the level of severity, and the type of harm. The three health-related categories are certainty or probability of cause; severity of hazard; and the type of harm. Statutory requirements for each of these vary substantially in the language used for each statute and often even within individual statutes. For example, the Toxic Substance Control Act

Table 9.1 RISK RATIONALE FOR DESIGNATION/REGULATION

Statutory Provision	Certainty/Causality/ Probability	Severity of Hazard (w/Risk)	Type of Harm
*TSCA § 4(a) (1)	(A) "may present" or (B) "is or will be" produced in substantial quantities and (i) "enters or may reasonably be anticipated to enter" or (ii) "there is or may be"	"an unreasonable risk of"	"injury to health or the environment" "the environment in substantial quantities" "significant or substantial human exposure"
TSCA § 4(f)	"there may be a reasonable basis to conclude that a chemical substance or mixture presents or will present"	"A significant risk of"	"serious, widespread harm to human beings from cancer, gene mutations, or birth defects"
TSCA § 5(b) (4) (A)	"presents or may present"	"an unreasonable risk of"	"injury to health or the environment"
TSCA § 5(f) [6(a)]	"there is reasonable basis to conclude" that it "presents or will present"	"an unreasonable risk of"	"injury to health or the environment"

Statutory Provision	Certainty/Causality/ Probability	Severity of Hazard (w/Risk)	Type of Harm
*CWA § 311(b) (2) (A)	"presents"	"an imminent and substantial danger"	"danger to the public health or welfare, including, but not limited to fish, shellfish, wildlife, shorelines, and beaches"

* [Toxic Substances Control Act, Ed.]
** [Clean Water Act, Ed.]

RISK RATIONALE FOR DESIGNATION/REGULATION

Statutory Provision	Certainty/Causality/ Probability	Severity of Hazard (w/Risk)	Type of Harm
CWA § 311(b) (4)	"may be"	"harmful to"	"the public health or welfare of the United States, including, but not limited to fish, shellfish, wildlife, and private property, shorelines and beaches"
CWA § 307(a)	"will cause"		"death, disease, behavioral abnormalities, cancer, mutations, physiological malfunctions in reproduction, or physical deformations in organisms or their offspring"
*SDWA § 1401(1)	"may have"		"any adverse effect on the health of persons"
SDWA § 1412(b)(1)(B)	"may have"		"any adverse effect on the health of persons"

Statutory Provision	Certainty/Causality/ Probability	Severity of Hazard (w/Risk)	Type of Harm
SDWA § 1421(a)(1),(d)	"may" result in and "may affect adversely"		"the presence of any contaminant" "the health of persons"
**MPRSA § 102(a)	"will not"	"unreasonably degrade or endanger"	"degrader endanger human health, welfare, or amenities, or the marine environment, ecological systems, or economic potentialities"

* [Safe Drinking Water Act, Ed.]
** [Marine Protection, Research, and Sanctuaries Act, Ed.]

RISK RATIONALE FOR DESIGNATION/REGULATION

Statutory Provision	Certainty/Causality/ Probability	Severity of Hazard (w/Risk)	Type of Harm
RCRA § 3001, § 1004(5)	(A) "may cause or contribute to" OR (B) "may pose"[when "improperly" managed]	"an increase in" "a substantial present or potential hazard to"	"mortality" or "serious irreversible, or incapacitating reversible, illness" "human health or the environment"
RCRA § 3002, § 3003, § 3004	"may be necessary to protect"		"human health and the environment"
FIFRA § 3(c)(5)(C)	(C)"[it] will perform its intended function without" [and (D)] "will not generally cause" [when used in accordance with widespread and commonly recognized practice" or as directed]		"unreasonable adverse effects on the environment"

* [Resource Conservation and Recovery Act, Ed.]
** [Federal Insecticide, Fungicide and Rodenticide Act, Ed.]

RISK RATIONALE FOR DESIGNATION/REGULATION

Statutory Provision	Certainty/Causality/ Probability	Severity of Hazard (w/Risk)	Type of Harm
FIFRA § 3(c)(7)(A)	"would not significantly increase"	"the risk of"	"unreasonable adverse effects on the environment"
FIFRA § 3(c)(7)(B)	"would not significantly increase"	"the risk of"	"any unreasonable adverse effects on the environment"
FIFRA § 3(c)(7)(C)	"will not cause"		"any adverse effects on the environment" during the time period covered by such registration
FIFRA § 3(d)(1)(B)	"will not generally cause" [same condition as below]		"unreasonable adverse effects on the environment"
FIFRA § 3(d)(1)(C)	"may generally cause" without additional regulatory restrictions in accordance with directions for use warnings, cautions, and for the uses for which it is registered ["when used in accordance with widespread and commonly recognized practice" or as directed] or when applied		"unreasonable adverse effects on the environment, including injury to the applicant"

Statutory Provision	Certainty/Causality/ Probability	Severity of Hazard (w/Risk)	Type of Harm
FIFRA § 6(b)	"generally causes" if it appears that a pesticide or its labeling or other material required to be submitted does not comply with the provisions of the Act or when used in accordance with widespread and commonly recognized practice		
FIFRA § 25(c)(3)	"in order to protect children and adults"		"[against] serious injury or illness resulting from accidental ingestion or contact"
*CAA § 108(a)(1)	"causes or contributes to air pollution which may reasonably be anticipated to"	"endangered public health or welfare"	

*[Clean Air Act, Ed.]

RISK RATIONALE FOR DESIGNATION/REGULATION

Statutory Provision	Certainty/Causality/ Probability	Severity of Hazard (w/Risk)	Type of Harm
CAA § 109	"are requisite to protect the public health"		
CAA § 111(b)(1)(A)	"causes, or contributes significantly to, air pollution which may reasonably be anticipated to"	"endangered public health or welfare"	
CAA § 112	"may reasonably be anticipated to result in"	"an increase in"	"mortality or serious irreversible, or incapacitating reversible, illness"
CAA § 157(b)	"may reasonably be anticipated to affect the stratosphere, especially ozone, if such effect" "may reasonably be anticipated to"	"endangered public health or welfare"	
CAA § 202(a)	"cause, or contribute to, air pollution which may reasonably be anticipated to"	"endangered public health or welfare"	
CAA § 211	"causes or contributes to air pollution which may reasonably be anticipated to"	"endanger the public health or welfare"	

RISK RATIONALE FOR DESIGNATION/REGULATION

Statutory Provision	Certainty/Causality/ Probability	Severity of Hazard (w/Risk)	Type of Harm
CAA	"causes or contributes to air pollution which may reasonably be anticipated to"	"endanger the public health or welfare"	
CERCLA § 102	"may present"["when released into the environment"]	"substantial danger to the public health or welfare or the environment"	
CERCLA § 104(a)(2)	"will or may reasonably be anticipated to cause" {"directly…or indirectly" after "release into the environment"]		"death, disease, behavioral abnormalities, cancer, genetic mutation, physiological malfunction including malfunctions in reproduction) (sic)" in "organisms or their offspring'

* [Comprehensive Environmental Response, Compensation and Liability Act, Ed.]

Statutory Provision	Certainty/Causality/ Probability	Severity of Hazard	Type of Harm
CERCLA 105(3)	not specified by statute	not specified by statute	
CERCLA 105(8)(A)	not specified by statute	not specified by statute	

Adapted from The 'Shotgun Wedding' of science and law: risk assessment and judicial review, *The Columbia Journal of Environmental Law*, 10, 67, 1985.

(TSCA) says "EPA may compile...a list of chemical substances that presents or may present an unreasonable risk of injury to health or the environment." It also says "EPA shall designate elements which present an imminent and substantial danger to the public health or welfare...." The Clean Water Act (CWA) says, "EPA may revise the list of designated toxic pollutants to include substances which will cause death, disease, behavioral abnormalities, cancer mutations...." Thus, elements of certainty and probability include "may," "will," "may affect adversely," "will not cause," "would not significantly increase," and others. The severity of hazard or risk includes "an unreasonable risk of," "a significant risk of," "an imminent and substantial danger," "harmful to," and "the risk of." Here, then, very early in the game, arose the new regulatory challenge: carrying out the mandate in the face of marginal or nonexistent data. How do we decide whether something "poses a risk," "will affect adversely," or "will cause harm," if there are no scientific data telling us one way or the other? Moreover, how do we distinguish between "may," and "will," or among "a significant risk," "harmful to," or an "imminent and substantial danger"? Once again, political demands create language that science, despite its ignorance, is expected to sort out. Davis described this as "basic science forcing."[3] Moreover, the language contains elements, not only of science, but of values in which personal points of view affect the perceptions dramatically.

In the absence of data, the debate became highly politicized very quickly. Many determinations were made by the courts as the EPA sued manufacturers and manufacturers sued the EPA, alleging overzealous regulation. It quickly became unfashionable to speak on behalf of chemicals; thus, the loudest voices insisted that the public could not wait for answers and that chemicals should generally be assumed guilty until proven innocent. That may, indeed, sound like both a laudable and simple solution to a very complex problem. Unfortunately, it is far from simple and, at times, is more harmful than helpful.

How did we get to this point in the early 1990s when regulatory risk language has demanded so much of the ill-prepared basic sciences?

THE GENESIS OF THE EPA'S FEDERAL CANCER POLICIES

One of the early debates regarding a chemical ban centered about the pesticide DDT. The substance of this debate illustrates both elements of scientific uncertainty and the political quality of the ultimate resolution. It clearly illustrates what William Ruckelshaus called "a shotgun wedding between science and law."[4] The chemicals have changed, as have some of the players, but identical debates continue today.

During this process of pesticide assessment, the EPA began developing its carcinogen assessment guidelines. The EPA attorneys summarized the various experts' positions put forth during these evaluations of carcinogenicity. The result was a series of cancer principles coming from a committee headed by Dr. Umberto Saffioti. The purpose was to address both the DDT issue and other issues raised by the Food Protection Committee of the Natural Resources Council of the National Academy of Sciences. Members of that committee had suggested that potential carcinogens might be added to foods if they were at "toxicologically insignificant" levels. They also suggested that safety might be assumed if the material had been in commercial use for a substantial period of time. Responding, in part to this and in part to the growing unrest over the general regulation of potential carcinogens, a 1970 report from an ad hoc committee of the Office of the Surgeon General enunciated a series of "guidelines" for the evaluation of potential carcinogens.

At the EPA DDT cancellation hearings, Dr. Saffiotti and his group cited these guidelines. Out of his testimony came seven general principles regarding carcinogenic hazards

1. Any substance shown conclusively to produce tumors in animals should be determined potentially carcinogenic in humans, except when the effect is caused by physical induction or where the route of administration is grossly inappropriate in terms of human exposure.

2. Carcinogenic data in humans are acceptable only when they present critically evaluated results of adequately conducted epidemiological studies.

3. No level of exposure to a chemical carcinogen should be considered toxicologically insignificant for humans.

4. Carcinogenic bioassays should include two species of animals of both sexes, with adequate control animals to subject lifetime administration of suitable doses, including highest tolerated dose, by routes of administration including those by which humans are exposed.

5. Negative results should be considered superseded by positive results, that should be deemed definitive, unless new evidence conclusively proves that the positive results were not due to exposure.

6. An implication of potential carcinogenicity should be drawn from both those tests that induce benign tumors and those resulting in tumors more obviously malignant.

7. Principles of zero tolerance are valid and should be expanded.[5]

In a subsequent hearing involving the ban of dieldrin and aldrin, a number of new principles were enunciated in addition to the foregoing seven. These were

1. The majority of human cancers are caused by avoidable exposure to carcinogens.

2. While chemicals can be carcinogenic agents, only a small percentage actually are.

3. There is great variation in individual susceptibility to carcinogens.

4. Carcinogenesis is characterized by its irreversibility and its long latency period following the initial exposure to the carcinogenic agent.

Shortly thereafter, Saffiotti attempted to codify formally as "officially noticed facts" 17 principles of carcinogenesis. After enormous and furious protests, these were reduced to "three basic facts." These related to the lack of a threshold, to the relationship between animal studies and humans studies and to the assumption that benign tumors were surrogates for malignant ones.

The principles, enunciated and fought over, represented a highly charged, politically efficient admixture of pure science and pure policy. That there is no threshold of exposure below which a carcinogen is not dangerous, that negative studies should be superseded by positive ones, that benign tumors equate to malignant ones, even that animals are meaningful surrogates for man were purely philosophical, policy-derived constructs. The concepts of variations in susceptibility, irreversibility, latency, and of the prevalence of carcinogenic properties of chemicals were, by contrast, scientific observations readily subject to experimental observation. Some of those we now know to be overstated or simply untrue. There are, for example, reparative mechanisms that reverse precancerous genetic changes.

These constructs set the stage for all cancer-related regulatory activity and continue to do so today. Furthermore, by being so firmly described as principles, they gathered about them an aura of certainty and scientific relevance that was clearly undeserved. This image has been nearly impossible to alter despite newer understanding of mechanisms of carcinogenesis, animal-human relationships, toxicokinetics, and other fundamental scientific principles that have proven many of those principles to be untrue, nonuniversal, or taken completely out of context. These newly enunciated cancer principles were first applied to the banning of the pesticide DDT. Responding to intense pressure from environmental groups with support from Dr. Saffiotti and others from the activist scientific community, the EPA issued a precancellation order on January 15, 1971.

Following the issuance of this document, a public hearing took place. In April 1972, Hearing Examiner Sweeney submitted to EPA his "Recommended Findings, Conclusions and Orders," which denied the ban on DDT. The carefully reasoned discussion in his 114-page opinion reveals both an understanding of the distinction between public policy considerations and scientific knowledge, as well as the requirements needed to establish scientific fact and scientific proof. He clearly recognized that, as we noted before, it is impossible to prove the negative, carcinogenicity, mutagenicity, and teratogenicity are, therefore, often beyond the present scientific state of the art. He understood well the substantial debate in the scientific community over how to relate laboratory results on small numbers of test animals at high dosage levels to low-level, long-term human exposure. His conclusions of law included:

> DDT is not a carcinogenic hazard to man. DDT is not a mutagenic or teratogenic hazard to man (at) the present dosages to which we are exposed. The Petitioners have met fully their burden of proof."[6]

Fully admitting DDT could not be regarded as either a proven danger as a carcinogenic nor as an assuredly safe pesticide, he opined that, "this evidence presented demonstrates a continuing need to pursue the truth as to the fact of DDT as a carcinogen for humans."

The hue and cry that followed produced intense political pressure and highlighted the need to set a framework for regulatory cancer policy. Ruckleshaus, director of the EPA, set that framework by overruling the hearing examiner's decision and finally banning DDT in June 1972. He adopted the principles noted above in applying the ban, stating that

DDT is a potential human carcinogen.

a. Experiments demonstrate that DDT causes tumors in laboratory animals;
b. There is some indication of metastasis of tumors attributed to exposure of animals to DDT in the laboratory;
c. Responsible scientists believe that tumor induction in mice is a valid warning of possible carcinogenic properties;
d. There is no adequate negative experimental studies in other mammalian species;
e. There is no adequate human epidemiological data on the carcinogenicity of DDT, nor is it likely that it can be obtained; and
f. Not all chemicals show the same tumorigenic properties in laboratory tests on animals.[7]

These statements illustrate what Ruckelshaus called a "shotgun wedding between science and the law." To support the policy of the DDT ban, decisions had to be made in the face of significant scientific uncertainty. These assumptions did not, with the stroke of a pen, fill in the scientific gaps and convert DDT into a known human carcinogen. What they did do was to permit the Agency to ban a chemical, as a matter of "prudent public policy," on the basis of limited or uncertain scientific data.

Of interest here is not the merit of the DDT ban, but rather the principles of our cancer policies. These principles fill in the void between scientific knowledge and the demands of health-related regulations, particularly those that focus upon risk, long latency diseases and low levels of chemical exposure. Now, nearly twenty years later, most Americans and even many scientists are unaware of the distinctions between scientifically established truths and presumptions accepted for regulatory purposes. Key among those is that chemicals that produce cancers in animals are "carcinogens" for humans as well. While undoubtedly that is sometimes true, it is certainly not a scientific law or principle. That argument has at least as much to deny it as it has to support it. The argument is a regulatory construct designed to bridge gaps in scientific knowledge. It lies in the field that might be called regulatory toxicology and should be distinguished from scientific toxicology. Certain principles of regulatory toxicology are contrasted with elements of scientific toxicology in Table 9.2.

Table 9.2 Regulatory Toxicology Versus Scientific Toxicology

Scientific Toxicology	Regulatory Toxicology
Generates and interprets scientific data	Develops policies
Studies responses of biological systems to toxic agents	Weighs risks versus benefits
Based upon experimentation	Makes safety judgments
Characterized by reproducibility	Responds to public and political demands
Apolitical	Varies from time to time
	Heavily reliant upon assumptions: • animal to human • high to low dose, and so on

The former incorporates prudent public policy presumptions into its methodologies. The latter deals with direct generation and interpretation of scientific data.

Scientific toxicology is a pursuit of the laboratory scientist who follows rigorous methodology, testing, and retesting in the manner described in Chapter 3. Regulatory toxicology is an admixture. It takes available data from the scientific arena and weighs and interprets those data. It also incorporates methodologies designed to overcome scientific uncertainties. The purpose of scientific toxicology is to uncover truths. The purpose of regulatory toxicology is to make decisions. Because the methodologies of regulatory toxicology are creatures of public policy, not of science, they vary in nature and are at the whim of the currents of public policy. They are directly influenced and molded by popular perceptions. Thus, a key difference between scientific toxicology and regulatory toxicology is that scientific toxicology is based upon the immutable principles inherent in the scientific method. Regulatory toxicology, by contrast, is an ever-changing admixture of methodologies intended to protect in the presence of uncertainties and to assuage public perceptions and respond to political pressures.

OCCUPATIONAL REGULATION

Occupational exposures present situations in which health risk controls may compete directly with clear and meaningful economic exigencies. There is, therefore, some historical difference within health-related regulation and legislation between that directed toward the occupational setting and that directed toward the environment in general. The workplace is sometimes a source of concentrated chemicals and other potentially injurious substances—asbestos, silica, radiation, coal dust, and the like. Occupational regulation has, however, been more resistant to the extraordinary demands of risk elimination that are so characteristic of the environmental regulation and legislation; however, the two are beginning to merge. The primary reason is fairly clear. Manufacturing facilities have no choice but to use chemicals and other agents and, therefore, to expose workers to some potential health risks. Now, with better industrial techniques and worker protection methodologies, the risks are much reduced. But attempting to eliminate these altogether often imposes unacceptable economic hardship on the industries. Here, very clear risks to the survival of a company or an entire industry and the jobs of its workers compete in a politically meaningful way with the ideal benefit of absolute protection of all workers. Occupational risk minimization is therefore generally less stringent than its environmental counterpart. Even here arguments rage and lines are often unclear: employers generally predict economic ruin and massive layoffs should proposed exposure limits be reduced; opponents claim that this is "nonsense" and point to historical evidence of similarly unfounded cries of "wolf."

For the most part, environmental limits imposed by both federal and state environmental protection agencies are far more stringent than are occupational controls. Constituents arguing for relaxed environmental standards (more "rational" or more "dangerous," depending on one's point of view) are no match for those out to save the planet and humanity from the destruction of our air and water. It is useful to compare permissible exposure levels in the workplace with those permitted by the EPA under provisions of the Clean Air and Water Acts. See Table 9.3 for some examples. It is even more instructive to calculate likely dosages in individuals exposed to these amounts either from working or from general environmental exposures.

Some would argue that these gross disparities indicate that industry doesn't care about the health of its workers; yet studies of workers are routinely done and exposure limits are modified in accordance with new data. Others would claim that EPA's zeal is so manifest that it imposes standards that stretch beyond all limits of rationality. Either of these positions is a matter of philosophy rather than of scientific certitude. Protection based upon clearly established scientific observations engenders little debate, for when we know that something is harmful at a given level, we are unanimous in our demand for protection.

Table 9.3 Comparison of Chemical Standards

Chemical Name	OSHA Standard	EPA Standard
Naphthalene	10 ppm TWA; 15 ppm 15-minute STEL*	0.004 mg/kg/day RfD**; 0.1 mg/l DWEL+
Xylene	100 ppm TWA; 150 ppm STEL*	2 mg/kg/day RfD**; 60 mg/l DWEL+
Tetrachloroethylene	25 ppm TWA*	0.01 mg/kg/day RfD**; 0.5 mg/l DWEL+
Trichloroethylene	50 ppm TWA; 200 ppm STEL*	0.007 mg/kg/day RfD**; 0.05 mg/l DWEL+
Toluene	100 ppm TWA; 150 ppm STEL*	0.2 mg/kg/day RfD**; 7 mg/l DWEL+

* TWA denotes 8-hour time weighted average
STEL is short-term exposure level (15 minutes unless otherwise noted)

** RfD is EPA Risk Reference Dose; denotes the estimated daily exposure to the human population, which appears to be without appreciable risk of deleterious noncarcinogenic effects over a lifetime of exposure

\+ Drinking Water Equivalent Level (DWEL) is the lifetime exposure concentration protective of adverse, noncancer health effects and assumes all of the exposure to a contaminant is from a drinking water source

Sources: American Conference of Governmental Industrial Hygienists. *Threshold Limit Values for Chemical Substances and Physical Agents,* 1991-1992. ACGIH, Cincinnati, 1991; U.S. Department of Labor. Occupational Safety and Health Administration; "Limits for air contaminants," 29 CFR 1910 (7/1/90 edition); U.S. Environmental Protection Agency, Office of Drinking Water, *Drinking Water Regulations and Health Advisories,* EPA, Washington, DC; EPA, 1991.

REGULATORY RISKS VERSUS REGULATORY BENEFITS

With the explosive growth of regulatory activity has come intense debate over how far to go, each side offering compelling and forceful arguments in support of its position. At times there is a conflict, either real or proffered, between human health and the health of other co-inhabitants of our planet. Macroeconomic and social theorists argue that environmental protection has profound adverse human effects: on jobs, standards of living, availability and quality of food in Third World countries, and, as a result, upon our health.[8] Those arguments, though sometimes substantive, are too remote generally to garner much support. How can a politician consider permitting cancers in exchange for food? Medical and scientific tradeoffs have not, therefore, favored chemicals or anything to do with them. The voices of chemical opponents have held center stage because their message is politically and popularly attractive. Other philosophical-scientific debates have swayed toward the sanctity of science when scientific arguments were consonant with political expediency. This occurs when arguments are more direct and poignant, as in the case of animal experimentation. The opinions of those who would regulate animal studies clash with the opinions of those who have benefitted from them. Diabetics, cancer victims, and survivors of devastating infectious diseases parade before congressional subcommittees, regaling members with personal tales of victorious fights against death or disability, pointing to animal studies as the sole reason for their survival. Here the rights of animals are pitted against the rights of those people. These debates raise a critical value judgment: do we sacrifice animals to save human beings? Only the most ardent of animal rights activists claim coequal status for rats and humans. And the political momentum can hardly ignore personal anecdotes of successful struggles against death. These debates demand difficult value judgments, made easier by the personal tales of individual victims. They also test the credibility of experts on both sides. Those supporting total bans on animal experimentation claim that new technology permits us to substitute computers for animals. Opponents claim that such arguments are bogus.

Politicians have some difficulty weighing the relative expertise, but they have little difficulty believing victims. Thus, animal experimentation persists.

Drugs, too, have tested the strength of opposing political wills. AIDS activists want drugs approved quickly, with minimal testing. They claim that the FDA drags its feet, insisting upon perfect safety: a silly requirement for a drug designed to treat a fatal disease. The FDA, on the other side, points to the greater public good served by cautious evaluation. They hark back to the days of snake oil salesmen and to recent experiences with useless or dangerous drugs such as Laetrile, Krobiacin, and Thalidomide.

The regulation of industrial chemicals presents quite the opposite political picture. Here, "victims" of Love Canal, Times Beach, Agent Orange, and other perceived chemically induced cancer epidemics are the constituents of anecdotal tales. No one comes forward to claim that small amounts of PCBs in the fish he or she ate cured his cancer or prevented his heart disease. There are no supportive popular testimonials. Thus, the credibility of opposing experts in these debates swings naturally and understandably toward indictment of these "poisons," to demand for cleanup and to more widespread future vigilance.

Thus, what is quite clear is that health-related environmental regulation represents a balancing of certain interests. If politicians are more persuaded that the public good, or more precisely, constituent interests, are served through imposed controls, then controls will be imposed. Their's is rarely a true scientific or medical risk benefit decision; more often than not, it is a political or risk/benefit decision.

SUMMARY

The agencies responsible for environmental protection march to the drums of popular demand. The substance of that demand is straightforward and quite consistent. No amount of any hazardous substance should be present in our air, our food, or our water. That principle seems unimpeachable and irrefutable, but when examined for scientific meaning, it makes little sense. As we discussed in Chapter 8, there is no such thing as "none." As instrumentation becomes increasingly sensitive, we find some of nearly every known substance in some level in air, food, and water. These may be removed so that none is found today, but tomorrow more will be found as measuring devices become more sensitive. Second, the term "hazardous," as we discussed earlier, also has little meaning when taken out of context. Nearly everything can be hazardous at some time and under certain conditions, and even things that are widely regarded as beneficial—fruits, vegetables, dairy products, and meats—contain small levels of natural carcinogens as part of their biochemical makeup. Thus, we cannot remove all chemicals and we cannot remove all potential hazards. The rise of environmental regulation has been a constant struggle to make sense of uncertainties and to harmonize inherent conflicts.

REFERENCES

1. Zimmerman, R. *Governmental Management of Chemical Risk: Regulatory Process for Environmental Health*, Lewis Publishers, Chelsea, MI, 1990.

2. Page, T., A generic view of toxic chemicals and similar risks, *Ecology Law Quarterly*, 7, 207, 1978.

3. Davis, D.L., The shotgun wedding of science and law: risk assessment and judicial review, *Columbia J. Environ. Law*, 10, 67, 1985.

4. Ruckelshaus, W.D., Science, risk and public policy, *Science*, 221, 1026, 1983.

5. DDT hearings: opinion and order of the administrator, Environmental Protection Agency, *Fed. Reg.*, 37, 13369, 1972.

6. Sweeney, E.M. Hearing Examiner's Recommended Findings, Conclusions, and Orders. Consolidated DDT Hearings (40 CFR 164.32). Environmental Protection Agency, pp. 113, Washington, D.C. (April 25, 1972).

7. Ruckelshaus, W., Opinion of the Administrator, Consolidated DDT Hearings, Environmental Protection Agency, p. 40. Washington, D.C. (June 2, 1972).

8. Wildofsky, A., Playing it safe is dangerous, *Regul. Toxicol. Pharmacol.*, 8, 283, 1988.

10
QUANTITATIVE RISK ASSESSMENT: AN ATTEMPT TO LINK SCIENCE TO CANCER POLICY

An understandable tension arises when political demands for health and risk-based controls are not met simply through scientific answers. "Why can't scientists just find out once and for all whether magnetic fields cause leukemia, or how much asbestos is harmful?" people want to know. The fact is that demands for answers do exceed actual answers; thus, domestic tranquility demands action despite incomplete data.

Mathematical risk assessment is a technique by which certain biological data can be converted into regulatory action. By the late 1970s pressures to regulate carcinogens were high indeed. Often the data upon which such regulation was based consisted solely of animal carcinogenicity studies. Invariably these studies tested large doses of the chemical, far higher than humans could ever experience from environmental exposures. To allow scientists to make judgments about low-dose human exposure from that high-dose animal data, mathematical formulas were developed. This mathematical extrapolation process became the field of quantitative risk assessment.

By using these mathematical formulae, the agencies were able to codify their approaches, thereby forestalling accusations of arbitrary decision making and explaining in some relatively consistent fashion their basis for decisions. Quantitative risk assessment is not designed to reflect scientific truths; it is designed to bring a semblance of order to chaos. Through quantitative risk assessment the regulatory agencies ease the tension between popular demands and scientific uncertainties. For the most part risk assessment is used to regulate carcinogens, though other long-term chronic health effects are considered, but far less thoroughly. It helps agencies make health-related decisions about chemical and other exposures in the absence of direct empirical data. It also attempts to bring stability, organization, and consistency into a system that can otherwise be chaotic, random, arbitrary, and capricious. It by no means is uniformly successful, but some form of formalized risk assessment is probably better than none at all. When an act asks for "ample margins of safety" or that "imminent dangers to health" be avoided, risk assessment permits us to convert "ample" and "adequate" to permissible exposure levels in the workplace, in drinking water, in air, and in food. It also permits the agencies to demonstrate mathematically how they arrived at those numbers. They weren't simply pulled out of a hat. Even with codified risk assessment methodologies, arguments rage: which model to use, how to interpret the data, what data to include, and what to exclude are but a few. But, employing these methods, agencies can make decisions and achieve some consistency. Furthermore, over the years the methods for developing rule assessment programs have been refined; some basic ground rules have been established. Opponents argue, with justification, that the rules are still capricious and uncertain and that each agency and each new administration deals with the matter differently. This is true, but at least it provides some structure, albeit a somewhat wobbly one. They also argue that risk assessment assumptions and models are arbitrary and capricious—a sham or a flim flam game—using mathematical tricks to obscure guesswork. This, too, has an element of truth. But, at least with this method, the rules are reproducible, public, and subject to scrutiny. Risk assessment, then, is an instrument of social compromise, providing numerical answers in the face of vast scientific uncertainties.

What it is not is science. No matter how many models are devised (or divined), no matter how sophisticated the mathematics or the statistics become, no matter how large we build the computers used to make computations; risk assessment methods and their results are limited by the quality of biological knowledge. Those steeped in the field of risk assessment, particularly mathematicians and statisticians, and also some environmental scientists, sometimes forget this. So involved are they with the process of their art, so committed to its outcomes, that they at times mistake further refinements in mathematical modeling for further refinements in knowledge. Mathematical modeling can be used to describe known biological behavior, but it cannot be used to predict biological behavior that is scientifically unknown. Thus, gaps in knowledge about the behavior of human subcellular and whole organism systems render mathematical descriptions no more than abject speculation. The fact that one develops mathematical models to describe cancer causation, does not mean that the models actually describe such behavior or that we have unlocked, through mathematics, the mysteries of carcinogenesis. Cancer causation is neither simple nor yet subject to established mathematical rules. We might analogize this to describing personalities by finding the right mathematical formulae to do so, but I think cancer is even more complex than personality. Some risk assessment modelers believe that the accuracy of the descriptions will be enhanced if the models are made more sophisticated. The fact is that making the models more complex and more sophisticated does not make the results any more biologically meaningful. Mathematical models that replicate cancer induction may, in fact, be developed, but those models must follow rather than precede an understanding of the processes. An example of what risk assessment is, how it is used, and a few of its underlying assumptions may help to explain why this is so.

In her outstanding book *The Dose Makes the Poison*, Alice Ottoboni discusses an example used by Dr. Alvin Weinberg.[1-2] Quoting Dr. Weinberg, she asks us to consider the biological effects of low-level radiation:

> Let us consider the biological effects of low-level radiation insults to the environment, in particular the genetic effects of low-levels of radiation on mice. Experiments performed at high radiation levels show that the dose required to double spontaneous mutation rate in mice is 30 roentgens of X-rays. Thus, if the genetic response to radiation is linear, then a dose of 150 milliroentgens would increase the spontaneous mutation rate in mice by 1/2 percent...Now, to determined at the 95 per cent confidence level by direct experiment whether 150 milliroentgens will increase the mutation rate by 1/2 per cent, requires about 8,000,000,000 mice! Of course, this number falls if one reduces the confidence level: at 60 percent confidence level, the number is 195,000,000. Nevertheless, the number is so staggeringly large that, as a practical matter, the question is unanswerable through direct scientific investigation.

Dr. Weinberg called this area that science is experimentally unable to address "trans-science." A good deal of risk assessment methodology operates in this region of trans-science, systematically providing answers to unanswerable questions.

The example of radiation is an instructive one because, despite certain known biological facts, it still leaves uncertainties. We know that radiation can cause cancer in both humans and animals. Yet, despite this near biological certainty, we cannot be certain how much it takes. We can make judgments, but these must be based upon assumptions in the absence of data. We cannot test the 8,000,000,000 mice needed to prove experimentally that low-level extrapolations, made from high-dose experiments are biologically accurate. We therefore make judgements, decide which kinds of models to use for our extrapolations and accept them for their intended purpose. We must never make the mistake, however, of assuming that these extrapolations reflect biological fact. It is just as likely, for example, that no mice will ever get cancer when dosed with 150 mg. Reparative mechanisms and other biological factors may render such a dose harmless, unable to produce cancer at all.

Note here that Dr. Weinberg considered in this example the number of mice needed to assess a 1/2% mutation rate, a rather large rate indeed by regulatory standards. The agencies have adopted as their index risk levels of 10^{-5} or 10^{-6}, not 0.5 such as we see in this example. Obviously, direct testing of such extraordinarily minute levels is completely out of the question. And that fact makes risk assessment the only way to reach numerical risk conclusions, but it does not increase its accuracy.

One of the most profound arguments between proponents and opponents of the risk assessment process as it currently rages is the question of thresholds.[3] That is, might a chemical or other agent (radiation, radon, asbestos) that produces cancer at a high dose in experimental animals, not produce it at all, ever, under the low-dose circumstances typical of environmental exposures. As are other issues of trans-science, this is one that can never be proved or disproved, merely hypothesized about, for or against. To be sure, as the quality of scientific data and general theories of carcinogenesis become better solidified, arguments can become more compelling, more consistent with basic theories of biological behavior. Thus, certain dose-response curves seen in epigenetic carcinogens (promoters) argue persuasively for threshold notions in those. Similarly, there are compelling arguments for thresholds among genotoxic agents as well. I have no intention of debating here the threshold/no-threshold issue. The fact that I want to emphasize, is that, in the absence of conclusive proof to the contrary (proof that, as we see from Weinberg's radiation discussion cannot ever be provided), the agencies utilize no-threshold concepts in assessing risks to people. Thus, all commonly accepted and utilized risk assessment approaches have built into them the public policy notion of no threshold. And, because that notion is not known to be true, all generally accepted models, no matter how sophisticated, are more mathematical reflections of public policy than they are of scientific fact.

The rejection of threshold is illustrated by the following example shown in Figure 10.1. If 1,000,000 people jump from a building 300 feet tall, 900,000 will die. Assume that the height of the building represents the dosage of the chemical. The very high doses (or tall buildings) represent the animal tests customarily performed. Now, to make judgments about effects of potential low doses, it is assumed that for each one-tenth the height is reduced, the number of people that die will be reduced by one-tenth. Thus, at 30 feet, it is assumed that 90,000 will die. At 3 feet, it is assumed that 9,000 will die and so forth. Clearly, this assumption is incorrect, because once we reach a certain relatively low height, say 0.3 feet (3.5 inches), no one will die, even though this model predicts 900

LINEAR EXTRAPOLATION IN
QUANTITATIVE RISK ASSESSMENT

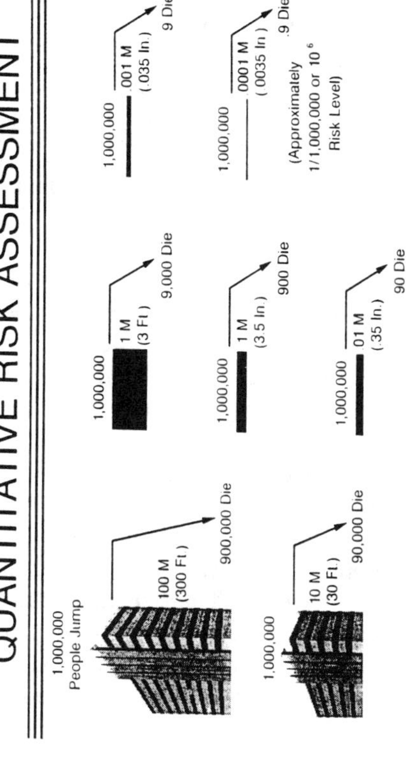
1,000,000
People Jump
100 M
(300 Ft.)
900,000 Die
1,000,000
10 M
(30 Ft.)
90,000 Die
1,000,000
1 M
(3 Ft.)
9,000 Die
1,000,000
1 M
(3.5 In.)
900 Die
1,000,000
.01 M
(.35 In.)
90 Die
1,000,000
.001 M
(.035 In.)
9 Die
1,000,000
.0001 M
(.0035 In.)
.9 Die
(Approximately
1/1,000,000 or 10^6
Risk Level)

Figure 10.1

deaths and continues to predict some deaths (90) even at a jump of .035 inches. The regulation of dioxin and other chemicals follow this model. It assumes that there is no dose that does not cause some people to develop cancer. This model may provide assurance of ultimate safety, but it undoubtedly overstates the risk of environmental background levels of dioxin.

When we examine some of the other assumptions inherent in risk assessment, the process moves even farther away from an accurate, proven reflection of biological behavior. In Dr. Weinberg's mouse example, we were dealing with both a known carcinogen and the very same species in which the carcinogenicity had been established. In most chemical regulation, we focus upon agents that are not known to produce cancer in people, but only in experimental animals, and in a limited number of animal species or even sexes or breeds at that. Consider, for example, the chemical trichloroethylene (TCE) regulated by the EPA as a group B2 carcinogen with a drinking water maximum contaminant level of 5 ppb.

Animal studies involving TCE in different strains of rodent are widely variable. Without commenting upon the quality of these studies, in one reported study B6C3F1 mice and Charles River rats were exposed to TCE. Some of the mice had a statistically increased incidence of cancers—far more in male than in female mice. None of the rats did.[4] In a study of Sprague Dawley rats and Swiss mice, there was no increased tumor incidence in either group.[5] In another study, NMRI mice, Wistar rats, and Syrian hamsters were given TCE. Female mice had a statistically increased incidence of lymphomas. The other species and male mice were not so affected.[6] Many other studies have shown similar variability.[7,8] Thus, rats, mice, and hamsters are not even good predictors of each other's responses. Females and males respond differently from one another; one genetic strain differs from the next. Clearly, such data cannot predict with any scientific solidity the response of human beings to similar chemicals.

Yet people don't want TCE in their water, and justifiably so. Thus, the methods of risk assessment must be employed, despite these additional uncertainties about the scientific validity of the methodology. This further limitation—assessing the relevance to humans of varied and often contradictory animal studies—again diminishes the biological significance of such models. No matter how sophisticated the mathematics, it will not represent a biological truism if these animal-to-human extrapolations are untrue. Again, this uncertainty renders risk assessment more of an instrument of the social sciences than of the biological sciences.

These are just a few of the uncertainties underlying this process. There are many more, each one compounding the limitations of risk assessment as an accurate predictor of human biological behavior. Hallenbeck and Cunningham discuss other sources of error in quantitative risk assessment. These include

1. Poor definition of the experimental and/or risk group exposure, such as, concentration, duration, chemical species, pertinent routes, dose rate.

2. Use of experimental studies involving less-than-lifetime exposure or short observation periods.

3. Use of experimental studies involving inappropriate route of exposure.

4. Toxicant interaction in the experimental studies or the risk group due to complex daily patterns of movement.

5. Pharmacokinetic and metabolic difference between species.

6. Improper control groups.

7. Extrapolation of the experimental results into the very-low-dose range.

8. Differences between the experimental and risk groups regarding: age at first exposure, sex, confounding exposures, smoking habits, and such.[9]

Some of these pertain to the use of human-to-human extrapolations in risk assessment as well as animal-to-human. Following this quite correct listing of uncertainties, Hallenbeck and Cunningham note, "Risk assessment is in the early stages of development. There is much room for improvement via elimination of sources of uncertainty and error."[9] Presumably they were referring to further discovery of biological data to reduce sources of error; "better" modeling will never do so. Furthermore, because so many of these issues are centered in the abstract realm of trans-science, many of those uncertainties can never be resolved through experimentation. Ten thousand studies over the next 1,000 years will not answer unequivocally the question about thresholds. Thus, we must be content with recognizing that risk assessment is a tool of the social sciences which is only mildly laced with biological scientific knowledge.

I frankly believe that there is nothing wrong with this. Risk assessment is a necessary aid to public policy. It forces systematic and codified behavior. It demands that rules be made based upon uniform methodologies.

As a social science tool, risk assessment serves as an important function. But its role in society is now so great that the public must learn to understand what it is and what it is not. If its purposes are misunderstood, risk assessment is at best misleading and, at worst, an instrument of demagoguery.

"Fifty Thousand Children Will Get Cancer from Alar" read a headline in the *New York Times*. That number, derived from risk assessment modeling, was the product of the Natural Resource Defense Council's effort to draw instant national attention to pesticide residues in foods. And "in a carefully orchestrated PR campaign,"[10] although untrue, it captured the attention of every American. Reasonable people might disagree about the concerns over pesticide residues in foods and whether they pose a sufficient health risk to warrant their total removal. That question is one of personal philosophy, risk-benefit assessment (as well as it can be done), and, ultimately, societal choices.

Disagreements might also occur over how the animal data used in the Alar risk assessment model should be weighed. What is absolutely clear is that there is no scientific truth to the statement that Alar will cause cancer in people under any circumstances, or, more specifically, under circumstances associated with the eating of Alar-sprayed apples. There are simply too many uncertainties. Alar is not known to cause cancer in people. In addition to which all of the uncertainties quoted above from Hallenbeck and Cunningham apply. It was clearly a misuse and misrepresentation of risk assessment data to extrapolate from one mouse study to a certain number of children and to present the public with the concept that Alar would give America's children cancer. It was not true and the public was misled. What could have been said without misleading the public?

It could have been noted that Alar was found at certain levels in some small fraction of tested apples. It could have been said that in one mouse study treatment with high doses of Alar led to an increase in tumors. It could have been noted that the significance of this finding to people is unknown. It could have been noted that risk assessment, which has little direct scientific relevance to humans but helps us to assess a proposed social policy, tells us that Alar may pose a public health risk. It could have been noted further that taking a position of prudence in the absence of certainty represents the position of the NRDC which recommends a ban. Each of these statements represent the truth concerning Alar. The statement put forth to the public did not.

To those with a messianic zeal to eliminate all unknown risks, such a "watered down" statement would seem laughable. "We could never get any action. We would never get anything done if we provided such a message." That aim was formally acknowledged by Mr. David Fenton of Fenton Communications, the public relations firm hired by the NRDC to highlight the Alar scare. The idea was for the story to achieve a life of its own, and continue for weeks and months to affect policy and consumer habits.

I believe that people are able to deal with truths and to make rational health-related decisions. It underestimates the intelligence and rationality of the American people to believe otherwise and places the zealous social manipulators who present their versions of the truth in the position of sole information controllers. This end-justifying-the-means methodology is dangerous in a free society, particularly one that is faced with profound and very costly risk-benefit decisions in coming years.

Risk assessment has, unfortunately, become a common instrument of such distortions. Because it presents numbers it has the aura of factuality. Because it comes from the government, it is believed. Its underlying assumptions are rarely probed for accuracy.

The Office of Management and Budget recently reported on the regulatory program of the United States. They cautioned that "policy-makers must make decisions based in risk assessments in which scientific findings cannot be readily differentiated from embedded policy judgments. This policy environment makes it difficult to discern serious hazards from the trivial ones and distorts the ordering of the government's regulatory priorities."[11] Effective decision making in the future demands that the public be provided a clearer explanation of the risk assessment process.

REFERENCES

1. Ottoboni, A.M., *The Dose Makes the Poison*, Vincente Books, Berkeley, 1984.

2. Weinberg, A.M., Science and trans-science, *Minerva*, 10, 209, 1972.

3. Upston, A.C., Are there thresholds for carcinogenesis? The thorny problem of low-level exposure, *Ann. N.Y. Acad. Sci.*, 534, 863.

4. Bell, Z.G., Olson, K.H. and Benya, T.J., *Final Report of Audit Findings of the Manufacturing Chemists Association: Administered Trichloroethylene Chronic Inhalation Study*, Industrial Biotest Laboratories, Inc., Decatur, 1978.

5. Maltoni, C., Lefermine, G. and Cotti, C., *Experimental Research on Trichloroethylene Carcinogenesis*, Volume 5, Princeton Scientific Publishing, Princeton, 1986.

6. Henschler, D., Elasasser, M. and Romen, W., Carcinogenicity study of trichloroethylene, with and without epoxide stabilizers in mice, *J. Cancer Res. Clin. Oncol.*, 104, 149.

7. Agency for Toxic Substances and Disease Registry, *Toxicological Profile for Trichloroethylene*, National Technical Information Service, Springfield, VA, 1989.

8. U.S. Environmental Protection Agency, Office of Health and Environmental Assessment, *Addendum to the Health Assessment Document for Trichloroethylene, Updated Carcinogenicity Assessment for Trichloroethylene*, U.S. Government Printing Office, Washington, DC, 1987.

9. Hallenbeck, W.H. and Cunningham, C.M., *Quantitative Risk Assessment for Environmental & Occupational Health*, Lewis Publishers, Chelsea, MI, 1986.

10. How a PR firm executed the Alar scare, *Wall Street Journal*, p. A22, col. 4, October 3, 1989.

11. *The Regulatory Program of the United States Government (April 1, 1990 - March 31, 1991),* Office of Management and Budget.

11
LITIGATION

If toxic perceptions commonly depart from the science of toxicology, nowhere is that disparity more manifest than in the courtroom. There, the wider the gap between perceptions and science, the greater the financial rewards. The interest of claimants is best served, not through dispassionate analysis of the merits of their toxic claims—sticking to the science. Rather, the drama of the courtroom and the salesmanship needed to sway juries, demands the magnification of perceptions and the minimization, or outright distortion of science. The more injured a claimant appears, the more heinous the miscreance, the nastier the hazardous substance, the higher the jury award. Our advocacy system turns on emotional appeals and depends upon the amplification of commonly held misperceptions. It is neither well suited nor well designed to uncover scientific truths.

The evolution of lawsuits focusing upon injuries from toxic substances has paralleled the awareness generally of the influence of such substances and their regulation. Before 1970 such lawsuits were relatively uncommon. By 1980 they were prosecuted and defended by hundreds of specialized attorneys and they had their own distinct name "toxic torts." Less than 10 years later, toxic claims commonly involved multiple claimants, sometimes thousands; hence, the new label "mass torts." These claims are so large and so expensive that insurance companies and corporations have full-time staffs devoted to their management and supervision.

Just as was the case in the early days of regulation, in the early days of toxicological claims, the science was clearer than it is today. Heavily exposed coal miners and insulation workers, disabled or killed by lung diseases and cancers, were clearly the victims of industrial toxins. Their heavy exposures and medically apparent consequences were readily connected. Jury awards were commonly given to these victims, the causes of their diseases properly connected scientifically, as well as by jury perception, to the identified cause. This was akin to the early days of environmental regulation when multicolored effluents in streams killed fish and had to be cleaned up. Then, the science of toxic effluents was matched quite directly by science-based regulatory responses. As time has passed, however, regulation has focused upon levels and qualities of chemicals for which the science of toxicity is less and less clear (see Chapters 7, 9, and 10). At the same time, popular perception of chemical-illness relationships and the enormous growth of the industry established to prove these relationships had resulted in litigation in which relationships between exposures and diseases are increasingly tenuous. Now, in toxic litigation, popular perceptions commonly hold sway over toxicological realities.

The litigation over Agent Orange may have been the most influential piece of toxic litigation. Its publicity contributed immensely to the awareness of chemicals as a potential cause of illness and of derivative legal rights of victims. At the same time, it stands as a prime example of the gulf between perception and science in toxic litigation. Most people believe today that our Vietnam veterans were poisoned by Agent Orange and that their poisoning produced a variety of diseases, particularly cancers in them and birth defects in their offspring. The scientific studies of Vietnam veterans and the reviews of specific claims of perceived victims of Agent Orange tell quite a different story. Twenty years, hundreds of millions of dollars, and 15 epidemiological investigations of Vietnam Veterans have produced no evidence that exposure to these defoliants gave rise to statistically meaningful increases in cancers or in birth defects in offsprings of soldiers (see Chapter 12).

Actual reviews of medical records of the veterans who claimed Agent Orange as the cause of disorders revealed a wide range of alternative explanations for their medical conditions. In one case, a lead claimant, who commonly appeared on television with his two children, each of whom had a malformation of the bones in the forearm, alleged that his exposure to Agent Orange was the cause. A review of his family's past medical records revealed that the deformities were handed down from generation to generation, a familial genetic disorder, long preceding Vietnam, and clearly unrelated to any exposures.

The Agent Orange experience stands as a stark reminder of the importance of perceptions to the outcome of toxic litigation. When the defendant companies conducted research surveys in numerous cities throughout the country, they attempted to assess the public perception about the relationship between Agent Orange and health effects. The results were loud and clear. People had made up their minds. Agent Orange had been adjudicated guilty in the public mind. They then attempted to learn whether minds could be changed if overwhelming scientific evidence were presented. The answer was no! The public would find liability against the manufacturers and give large awards to the veterans notwithstanding the evidence. Armed with this convincing data that the science had little chance of overcoming the perceptions, the companies settled the Agent Orange claims with a fund of $180 million.

The Agent Orange litigation was so prominent that it was both an effect of and a cause of toxic perceptions. Veterans sued because they believed they were poisoned. The public learned of their stories, which reinforced what they knew to be true about toxins and their devastating effects.

Thus, lawsuits are both a result and a cause of toxic perceptions. When people believe that they have been poisoned by a toxic chemical, true or not, they file lawsuits. When lawsuits are large in scale, or of general interest, they confirm public opinion, adding to perceptions of toxic hazards.

Litigation over toxic substances may have arisen in response to real and established threats. Today, however, it has grown into a billion-dollar industry, fueled by the widespread perception that toxic substances are threatening our health and welfare. The growth of toxic perceptions has been paralleled by a growth industry of toxic litigation, independent of scientifically definable causal connections. Such litigation comes from a variety of origins.

QUESTIONABLE CLAIMS GRAFTED ONTO A SETTING OF HISTORICAL TOXICITY

The legacy of serious asbestos toxicity is a sad page in American industrial history. People were reduced to dependency upon respirators and many died from asbestosis and cancer associated with the exposure. Claims cost companies and insurers billions of dollars. Companies declared bankruptcy; others are still on the verge.

As clear as the relationship between asbestos and human suffering was, the new page in this litigation history passed from the annals of science into the realm of perception. Now claims are being made by thousands of individuals with far lower exposures, sometimes with little or no diseases, often with little ability to separate such factors as cigarette smoking from these small asbestos exposure as potential causes of measurable lung disease.

These perceived illness-asbestos relationships come about in a number of ways and for a number of reasons. First of all, the notoriety of asbestos has primed people to consider asbestos as a culprit for any and all lung conditions, asbestos-related or not. People want explanations for their illnesses (other than smoking which suggests personal responsibility). Asbestos is a good target. The decline in blue-collar jobs has contributed to this trend. When people lose jobs, filings for health-related compensation increase. Unions encourage workers to seek causes such as asbestos. And, finally, attorneys encourage it. Many have made substantial livings and long professional careers dealing with the scientifically meritorious high-level asbestos disease claims. Their research and experts and systems are in place. It makes good business sense to continue in that line of work. The fact of weak science in the low-level exposure claims does not deter filing and proving claims.

Thus, even as levels of toxins decline, institutional memories persist. Toxins may become permanent subjects of litigation whether or not they are responsible for claimed illnesses.

CHEMICAL FEARS AS A CAUSE OF ILLNESS AND CLAIMS

In Chapter 2 we discussed the clinical ecologists and the patients for whom they care. For the most part these patients are a group of people who share common characteristics. They are afraid of chemicals and of our industrial life. They are suggestible. They are readily convinced by their clinical ecologist physicians that they have real, chemically-induced diseases. They are also commonly litigants, suing their employer or a company, claiming that they were poisoned by one or another chemical.

A family of four—mother, age 28, father, age 29 and children ages 7 and 8—moved from their suburban home to a log cabin in the backwoods of Mississippi and sued the company that had manufactured and the individuals who had applied a pesticide to their home. The parents believed that they and their children had been poisoned by this chemical, and, moreover, that this toxicity had rendered them vulnerable to other chemical assaults and unable to function in modern society. Before they fled to the woods to live as recluses, they received medical advice from a clinical ecologist who, among other things, had the children instruct their school about certain do's and don't's. This included what cleaning agents could and could not be used in the school, prohibitions against the children's use of paints and crayons, and other, unfounded and socially disruptive restrictions. Eventually, after fighting with school officials and others who were neither understanding nor able or willing to accommodate their needs, the parents removed their children and left civilization. Ironically, they moved into a cabin whose logs had been impregnated with pentachlorophenol and other chemicals associated with creosoting. Nonetheless, they felt that this new "chemically free" environment was salutary and their doctor concurred.

Nonetheless, they felt that this new "chemically free" environment was salutary and their doctor concurred.

The trial was a test of science versus perception. Not a shred of legitimate scientific evidence linked any physical ailments to the treatment of their home with the pesticide. The perceived link was made by the parents and confirmed by their nonmainstream physician whose personal theories were brought into the courtroom to prove the connection. In rendering a verdict against the claimants, the jury found against the perceptions of toxicity and in favor of the scientific evidence.

This result may have occurred for several reasons which distinguished the matter from the Agent Orange litigation. First, the position of the claimants and their doctor was extreme, beyond common concerns about chemicals. Second, the claimants were convinced that ordinary products such as crayons posed a risk, a notion that the jury could not buy. Third, there were contradictions. Despite a perceived intense chemical sensitivity, the family now lived in a home in which chemicals were more prevalent than before. In a way the jury was still swayed by perception, but in this instance its perception ran contrary to that of the claimants and coincided with scientific toxicology. In the case of Agent Orange, by contrast, popular perceptions strongly opposed the science and supported the claimants.

Another claim arose because of a chemical fear and a variety of unexplained health complaints that the claimant was certain were due to chemical poisoning. A 45-year-old man bought a chemical, pentachlorophenol (PCP), to treat wood pillars in his basement. He was well protected with appropriate equipment. That night his child vomited and the family took him to an emergency room. The illness was undoubtedly due to a standard stomach virus, but the man associated the event with the chemical smell and believed that the child was poisoned. From that point on, a variety of minor and standard illnesses in that family were, in the man's mind, due to the chemical. A month or so later, he was rushed to an emergency room complaining of difficulty breathing. The diagnosis was "hyperventilation," an anxiety-related condition. He then had the state health department visit his home. They found no problem but did let him know that PCP could cause "neurological injury." At that point he developed an anxiety-related tremor. Symptom after symptom developed and expanded, one on top of the other, until this man, convinced he was poisoned, became increasingly disabled. Every time he entered his home the symptoms intensified, until he moved out in self-defense. In an unsuccessful lawsuit the man claimed that he was poisoned recognized that his complaints were exaggerated: the result of fear and worry rather than of direct poisoning. Again the jury verdict supported the science.

These stories are repeated everyday. Lawsuits arise because people develop psychological responses to perceived poisonings. Others are brought by individuals who are truly ill and believe that their illnesses are due to a chemical poisoning. Many illnesses that lack readily identifiable causes are grist for such suits. Cancers, birth defects, multiple sclerosis, lupus erythematosus, scleroderma, and dozens of others are among them. Profound chemophobias, certain knowledge of the harm that chemicals cause and an abiding human drive to find causal explanations for diseases have made chemicals and their manufacturers the hands-down target of choice for such actions.

The outcomes of these cases commonly follow popular perceptions. If the claimant's beliefs match those of the jury, the claimant prevails. If the claimant seems off base, he or she loses. Science per se wins only when it supports the prevailing perceptions of jurors. Thus, toxic perceptions strongly determine the outcome of toxin-related lawsuits.

PUBLIC OUTRAGE AS A CAUSE

When people learn that their most precious commodities—their air and water—are fouled by noxious or hazardous chemicals, they are understandably outraged. That outrage, sometimes helped along by self-interested attorneys, frequently translates into lawsuits. Often those suits allege a wide variety of illnesses and disabilities, many of which (if not all) had nothing to do with the exposures in question. Often, in fact, actual personal exposures cannot even be established and certainly not to amounts of chemicals known to cause harm.

When the residents of the city of Tucson found that their water had been contaminated with trichloroethylene, a degreasing solvent used at the airport during the Second World War, nearly 2,000 residents sued the city, the airport, Hughes Aircraft, and the federal government. The allegations that they made read like a compendium of all human diseases and complaints. Everything from headaches to hemorrhoids to hernias to cancers and birth defects was blamed upon this chemical contamination. The fact is that science permitted no connection between any of these complaints or illnesses and this chemical, but perceptions were strong. In that instance, the sheer number of claimants with the potential for enormous damages combined with the understandable emotional impact that drinking water could be contaminated, drove a large out-of-court settlement. As in the Agent Orange matter, the defendants believed that they could not overcome the perceptions of an outraged jury. Scientific realities would not be persuasive.

ECONOMIC STRIFE AND ENTREPRENEURIAL ATTORNEYS

In the East Texas town of Dangerfield, a mass toxic tort claim is currently in progress that has been dubbed "the attorney employment act of 1989." When the Lone Star Steel Plant closed in Dangerfield, half the residents were thrown into unemployment. An entrepreneurial attorney in town used this economic strife to his (and the townspeople's) advantage filing lawsuits on behalf of nearly 2,000 former Lone Star employees. Like the claims in the mass toxic tort Tucson case, these claims alleged that every health problem ever visited upon these individuals was caused by toxins in the workplace. Unlike that claim, specific causes were not alleged. Rather, 400 defendants were named, ranging in scope from chemical companies to hardware stores. Any distributor or manufacturer ever dealing with the Lone Star Plant became a defendant.

How a jury would perceive this bizarre case is really not relevant since, by the time of trial, millions of dollars in settlement will already have been paid; many say "extorted." The town's only judge made a series of rulings early on blocking any dismissals of even the most untenable of defendants forcing them to remain in the case until the completion of pretrial discovery, some years away. To forestall further defense costs, many have settled out-of-court, an estimated $12 million dollars worth to date.

The Lone Star litigation exemplifies the most extreme excess of toxic litigation. It is a case completely devoid of scientific or legal merit, born solely of economic deprivation and attorney greed.

CLUSTERS

When clusters of diseases arise, particularly among children, both claimants and jurors readily assume that there must be a common cause. When those clusters are combined with an identified source of contamination, that source readily becomes the perceived cause. Such perceptions may be so profound that they prompt, among those who believe themselves victims, an intense zealotry. That happened around Love Canal.

Love Canal is the name most closely attached to hazardous wastes. This community near Niagara Falls, New York, discovered that it was built near and on a former hazardous waste landfill. Local residents quickly added up the numbers of cancers, miscarriages, and birth defects in the neighborhood and concluded that they were suffering an epidemic and that the chemicals were to blame. Homes were purchased, people moved, lawsuits filed and settled. The epidemiological studies never established a community-specific increase in those diseases.[1,2] These perceived "clusters" of birth defects and cancers led the residents to focus upon a common cause. Clusters are commonly observed random distributions of either apparent or real excesses of a common disorder. In many cases, as in Love Canal, clusters are only apparent. When a study is properly set up with controls and a study group (see Chapters 3 and 5), there is no actual increase. In other cases, clusters are statistically explainable phenomena, having nothing to do with a common cause. Clusters, however, drive toxic litigation. They appeal to human nature's quest for common and explainable causes of serious diseases.

ROLE OF EXPERTS AND JUNK SCIENCE

Any complex litigation, including toxic torts, requires expert testimony to support the claimants' assertions. Physicians and scientists must come forward and testify to the connection between the claimants' disorders and the claimed toxic cause. When exposures were great and disease clear, experts could make such connections clearly and honestly. Newer claims have tested the credibility of experts, forcing them to stretch and reach beyond the science of toxicology, into the realm of perceptions. Peter Huber, in his book *Galileo's Revenge: Junk Science in the Courtroom* has decried this practice viewing it as: "Scientific humbuggery in court...an immensely profitable business."[3]

Linking low-level environmental exposures to specific diseases in specific individuals commonly requires moving beyond any available scientific data. For a specific individual, how much was absorbed into the system, how well other potential causes of the disorder have been considered and excluded[4-11] and whether a risk is real or based upon regulatory assumptions[8,9,12-15] are essential to the scientific evaluation of cause and effect.

In claims of illness or risk associated with environmental contaminants, these required "facts" often provide insurmountable hurdles for claimants for they ask for measurements that are unavailable and studies that do not exist. Consider claims of cancer risk as an example. Nowhere is the merging of public and social policy with common law "truths" more apparent. The fact that an agent is regulated as a potential human carcinogen is regularly used to argue risk to a specific claimant. Quantitative risk assessment methodologies are even used to weigh that risk. The numerous assumptions and speculations underlying risk assessment have been discussed extensively in other papers[12,14,16-18] and in Chapter 10. Risk assessment recognizes the need to protect against perceived, possible, or uncertain risks. But in toxic tort claims in which a claimant must prove a future risk, a higher level of proof is expected.

Proving that current illnesses are due to environmental pollutants similarly often outstrips the availability of scientific facts. In the claims in Tucson, Arizona, the public's clamor for redress notwithstanding, modern science knows of no actual or potential harm produced by such levels of the chemicals found in the water and, certainly, no connection to the claimed impairments. Science cannot, in this case, provide the facts demanded by these claims.

The claimants' response to this factual vacuum has been twofold: to argue for changes in evidentiary requirements lessening the claimants' burden of proof, and to purchase testimony from a growing industry of "fact" manufacturers.

In scholarly treatises it has been argued that these difficulties in proof bespeak a need to change our standards of proof.[12,19-21] This argument recognizes the difficulties of the claimants' burdens, understands the limitations of science in providing answers, and honestly asks for a reassessment of the role of the court. Whether or not one agrees with this recommendation, at least it has the virtue of honestly recognizing and separating today's scientific capabilities from today's litigation demands.

Far more disturbing is another trend, a burgeoning of science-for-hire designed to provide an imprimatur for the courtroom in the guise of scientific knowledge. A growing contingent of physicians have departed from the empirical sciences to fashion this new medicine specifically molded to the demands of toxic tort cases. Tests, without proper study or controls are being introduced at an astounding rate claiming to show neurological dysfunction, immunological disorders, neuropsychological aberrancies, metabolic derangements, and other biological malfunctions. Commonly, these are solely tort tests, not used clinically to evaluate patients in actual health care, but only to provide data for claims. The blink test, sway test, certain measurements of pollutants in bodily tissues and fluids, and various antibody tests fall into this category.[22] Other tests may have a place in the research laboratory or in a few special clinical settings, but in the context of these claims they are being used inappropriately or interpreted incorrectly. In some instances these methodologies have been formally repudiated by the peer community.[23] In many cases, however, the scholarly medical academies and respected academicians are unaware of these activities and are shocked to learn of them. In essence, this testing and testimony industry has grown to meet claimants' demands for "proof," when proof is unavailable, to satisfy a system that insists that the claimant meet his evidentiary burden.

The purveyors of this courtroom science depend upon several factors for their success. One is the law of supply and demand. The second is the popular perception about the dangers of chemicals. The third is the simplistic, easily understood nature of the claimant's allegation. "I was exposed. I am sick or will be. Therefore, the chemical is responsible." Finally, they depend upon the fact that judges, jurors, and opposing counsel are scientifically and medically naive and, therefore, readily misled. To understand how this courtroom science is constructed and, how laymen are misled, a passing understanding of two disciplines is necessary. The first is the science of causation analysis. The second is the concept of controlled scientific studies. This latter discipline we have discussed thoroughly in Chapter 3.

Sir Bradford Hill (Chapter 5) discussed the epidemiological elements needed to establish the potential for causation. Methods for investigating causal relationships between agents and illnesses in specific individuals also do exist and are well described in the scientific literature.[4,7,24-31] Hill's methodology differs in that it asks the more global question about the capability of the substance to produce the disease under investigation. Individual causation methodology expands that concept and applies it to a particular individual.

To introduce the proper methodology for assessing causal relationships, let us begin by using examples from actual cases. These illustrate basic and easily understood flaws in customarily admitted testimony.

A witness testifies that a claimant's lung cancer was probably caused by exposure to toluene diisocyanate. The methodological flaw is that this chemical is not a known carcinogen. If a chemical is not known to cause cancer, then it is scientifically inaccurate to link it to a particular case of lung cancer. A scientifically well-founded conclusion about a cause and effect relationship must be supported by scientific evidence that such a relationship has been demonstrated and, therefore, can occur. Otherwise it is pure opinion playing into perceptions, defying scientific understanding.

An expert may also play into popular perception by the misleading equating of organ system disorders. People generally believe that such terms as "kidney injury," "liver disease," and "nerve damage" accurately reflect the consequences of chemical exposures. The fact is that there are major distinctions between types of kidney, lung, liver, and neurological diseases, some of which have and others of which have not been scientifically linked to chemical causes.

Most diseases have classical clinical and morphological (seen under the microscope if specimens are available) patterns. Diabetes causes a specific kind of kidney disease; certain chemical solvents produce quite another pattern. Carbon tetrachloride causes a characteristic liver injury; viral infections generally produce a different kind. The clinical course of viral encephalitis is generally different from that of chronic lead poisoning. Smoking-induced lung disease is different from asbestosis. The peripheral neuropathies associated with hexane are different from those associated with lead poisoning, alcoholism, or diabetes. Benzene is known to produce a certain kind of leukemia, but only that kind and no other cancer. Vinyl chloride is known to cause a rare cancer, hemangiosarcoma of the liver, not squamous cell carcinoma of the lung or carcinoma of the uterus.

Despite the fact that there exists a specificity of causal agent to disease process, testimony is routinely brought into the courtroom, which pretends that no such specificity exists. Formaldehyde can cause renal injury and therefore it caused this man's kidney failure opined an expert, despite the fact that the man had classical diabetic nephropathy and no features of a kidney disorder ever described in association with formaldehyde exposure.

The field of toxicology is founded upon the dose response principles discussed in Chapter 4. When an exposure to a chemical is less than that known to produce a toxic response, scientific data cannot, as a rule, support a claim of a causal connection. Despite this, experts commonly testify to causality notwithstanding exposures well below levels known to be injurious. This is especially true in the environmental claims in which ppm and even ppb are common units of exposure.

Expert testimony commonly occurs despite other analytical errors. An oncological surgeon provided opinion testimony that his patient's lung cancer was caused by exposure to a chemical in a paint product. The chemical is not a known carcinogen and the first exposure occurred only 4 months before metastatic lung cancer was diagnosed, far too soon to satisfy the known behavior of carcinogens.

The proper principles of the methodology of causation analysis can be summarized as follows:

Can: Can the agent in question produce the disease at issue?

1. Are there substantial and properly relevant animal data?
2. Is there human evidence, particularly epidemiological support?

Did: Did it cause the disease in this case?

1. Have other causes been properly considered and ruled out?
2. Has the exposure been confirmed?
3. Was the exposure sufficient in duration and concentration?
4. Was the clinical pattern appropriate?
5. Is the morphological pattern appropriate?
6. Is the temporal relationship appropriate?
7. Is the latency appropriate?

Just as proper medical practice requires a systematic consideration of the differential diagnoses, proper causation analysis requires the use of the methodology outlined above. The importance of this has been delineated by other authors beginning with Koch in the 19th century and continuing today among environmental pathologists and clinical scientists studying adverse drug reactions.

Once this approach to causation analysis is understood, a second question arises. What kinds of studies need to be conducted to tell us whether a chemical is capable of producing a disorder? The answer to this is found in our discussions in Chapters 2, 3, and 5 in which we considered scientific study design and Hill's criteria by which the weight of epidemiological data could finally be accepted as establishing a causal nexus.

To study properly the effect of an agent, we must know that the agent actually entered the subject's body, how much of it entered, and how that differed from the controls. Unless we know this, we cannot conclude that the test agent was the cause of a disorder.

Toxic tort testing, both laboratory and epidemiological, is commonly flawed in all of these respects. Consider these examples. In a mass tort claim it was alleged that a test known as the "blink" test was abnormal in a group who drank TCE-contaminated well water and that the TCE was the cause. Even if the "normal" response to the blink test were well-established, which it is not, and the residents' tests were abnormal, it would be literally impossible to point to the drinking water as the culprit.[22,32] Why? Because the levels actually drunk by individuals were unknown; studies had never established a relationship between low levels of ingested TCE and blink test abnormalities; and because at levels of ppm or ppb we are exposed to thousands of chemicals any of which could potentially be causal—thousands of confounding variables.

In an epidemiological study it was claimed that residents living near a site of air emissions (in the ppb range) developed a variety of complaints due to that exposure. The "controls" had no demonstrably different exposure level (there were no comparative measurements) and at the level measured those chemicals are just as high routinely in indoor household air.[33] Thus, conclusions about relationships between the exposure and symptoms were marred by an invalid control group and by failure to control the confounding variables.[34,35]

Immunological tests are commonly used to "prove" a toxic cause of immune dysfunction. Wide day-to-day variations in "normal" permits witnesses to claim abnormalities when a repeat test would disprove that. Frequently exposures are alleged, but not defined; therefore, the alleged exposure cannot be properly connected to a test result. Effects of the chemical in question upon that test are often unknown, thus conclusions are pure guesswork. Confounding variables are never controlled.

There are really only a few solutions to this growing gulf between the availability of valid science and the public's demand for answers and testimony in toxic tort claims. One is to formalize the proposition that the resolution of these claims are matters of policy, not truths. If, however, evidentiary requirements are not changed—and I, for one, believe that they should not be—then those of us who are alarmed by pseudo-truths masquerading as truths and by the distortion of science must help judges and juries to become better informed.

There is some indication that judges are becoming better informed and better able to protect juries from prejudicial or misleading testimony. There have been a number of recent cases in which "scientific" testimony has been excluded or overturned for lack of reliability. Some of those cases are documented.[36-40] Recent reviews of those decisions are available.[41-44] The courts face growing challenges, however, as these new "sciences" become increasingly sophisticated and espoused by well-credentialed witnesses.

SUMMARY

Claims of injuries from exposures to toxic substances are growing in frequency and magnitude. They are brought about by an admixture of toxic perceptions and other influences: outrage, fear, greed, and economics. They attempt to play upon the popular toxic perceptions of a jury. Often that favors the claimants, but not invariably. What is quite clear, however, is that perceptions more than toxicological realities determine the directions and outcomes of these cases. Medical and scientific experts, required to serve that claimant's cause must frequently stretch beyond the limits of scientific realities to serve the needs of their clients and play to the primed perceptions of a receptive jury.

REFERENCES

1. Grisham, J.W., Ed., *Health Aspects of the Disposal of Waste Chemicals*, Pergamon Press, New York, 1986.

2. Andelman, J.B. and Underhill, D.W., *Health Effects of Hazardous Waste Sites*, Lewis Publishers, Chelsea, MI, 1987.

3. Huber, P.W., *Galileo's Revenge: Junk Science in the Courtroom,* Basic Books, 1991.

4. Gots, R.E., Medical causation and expert testimony, *Regul. Toxicol. Pharmacol.*, 6, 95, 1986.

5. Evans, A.S., Causation and disease: the Henle-Koch postulates revisited, *Yale J. Biol. Med.*, 49, 175, 1976.

6. Hackney, J.D. and Linn, W.S., Koch's postulates updated: a potentially useful application to laboratory research and policy

analysis in environmental toxicology, *Am. Rev. Resp. Dis.*, 119, 849, 1979.

7. Irey, N.S., *Syllabus*, Division of Experimental Pathology, Armed Forces Institute of Pathology, Bethesda, MD, 1982.

8. Ruckelshaus, W.D., Science, risk, and public policy, *Science*, 221, 1026, 1983.

9. National Research Council, *Risk Assessment in the Federal Government: Managing the Process*, National Academy Press, Washington, DC, 1983.

10. Office of Technology Assessment, *Assessment of Technologies for Determining Cancer Risks from the Environment*, U.S. Government Printing Office, Washington, DC, 1981.

11. Wilson, R. and Crouch, E.A., Risk assessment and comparisons: an introduction, *Science*, 236, 267, 1987.

12. Schwartzbauer, E.J. and Shindell, S., Cancer and the adjudicative process: the interface of environmental protection and toxic tort law, *Am. J. Law Med.*, 14, 1, 1988.

13. Ruckelshaus, W.D., Risk, Science, and Democracy, *Issues Sci. Technol.*, 1, 19, 1985.

14. U.S. Environmental Protection Agency, *Risk Assessment and Management: Framework for Decision Making*, U.S. Environmental Protection Agency, Washington, DC, 1984.

15. Russell, M. and Gruber, M., Risk assessment in environmental policy-making, *Science*, 236, 286, 1987.

16. Perera, F.P., Quantitative risk assessment and cost-benefit analysis for carcinogens at EPA: a critique, *J. Public Health Policy*, 1987, 202, 1987.

17. Gleeson, J.G. and Gots, R.E., Understanding and defending claims of increased risk of contracting disease, *DRI Damages and Jury Persuasion*, 6, 54, 1987.

18. Davis, D.L., The shotgun wedding of science and the law: risk assessment and judicial review, *Columbia J. Environ. Law*, 10, 67, 1985.

19. Novey, L.B., Ed., *Causation and Financial Compensation for Claims of Personal Injury from Toxic Chemical Exposure: Conference Proceedings*, Novey, L.B., Ed., Georgetown University Institute for Health Policy Analysis, Washington, DC, 1986, 233.

20. Rosenberg, D., The causal connection in mass exposure cases: a public law vision of the tort system, *Harvard Law Rev.*, 97, 849, 1984.

21. Trauberman, J., Statutory reform of toxic torts: relieving legal, scientific and economic burdens on the chemical victim, *Harvard Environ. Law Rev.*, 7, 177, 1983.

22. Feldman, R.G., Chirico-Post, J. and Proctor, S.P., Blink reflex latency after exposure to trichloroethylene in well water, *Arch. Environ. Health*, 43, 143, 1988.

23. Terr, A., Clinical ecology, *Ann. Intern. Med.*, 111, 168, 1989.

24. Evans, A.S., Causation and disease: the Henle-Koch postulates revisited, *Yale J. Biol. Med.*, 49, 175, 1976.

25. Hackney, J.D. and Linn, W.S., Koch's postulates updated: a potentially useful application to laboratory research and policy analysis in environmental toxicology, *Am. Rev. Resp. Dis.*, 119, 849, 1979.

26. Irey, N.S., Tissue reactions to drugs, *Am. J. Pathol.*, 82, 617, 1976.

27. Irey, N.S., Diagnostic problems in drug-induced diseases, in *Drug-Induced Diseases*, Volume 4, Meyer, L.P., Ed., Excerpta Medica, Amsterdam, 1972.

28. Meyer, L. and Herxheimer, A., *Side Effects of Drugs*, Volume 7, Excerpta Medica, Amsterdam, 1972.

29. Henninger, G.R., Drug and chemical injury, in *Pathology*, 6th edition, Mosby, St. Louis, 1971, 174.

30. Gots, R.E., Medical/Scientific Decision Making in Occupational Disease Compensation: Analytical System, Operational Approach, presented at Crum and Forster Corporation, November 1, 1981.

31. Gots, R.E., The science of medical causation: its application in toxic tort litigation, *DRI Toxic Tort Litigation*, 4, 18, 1986.

32. Feldman, R.G., Firnhaber, W.R., Currie, N.N., et al., Long-term follow-up after single toxic exposure to trichloroethylene, *Am. J. Ind. Med.*, 8, 119, 1985.

33. U.S. Environmental Protection Agency, *The Total Exposure Assessment Methodology (TEAM) Study*, National Technical Information Service, Springfield, VA, 1987.

34. Ozonoff, D., Colten, M.E., Cupples, A., et al., Health problems reported by residents of a neighborhood contaminated by a hazardous facility, *Am. J. Ind. Med.*, 5, 581, 1977.

35. Ozonoff, D., Medical aspects of the hazardous waste problem, *Am. J. Forensic Med. Pathol.*, 3, 343, 1982.

36. *Richardson v. Richardson-Merrell,* 649 F. Supp. 799 (DDC, 1986).

37. *Sterling v. Velsicol Chemical Corporation,* 647 F. Supp. 303 (W.D. Tenn. 1986).

38. *In re Agent Orange Product Liability Litigation,* 611 F. Supp. 1223 (E.D.N.Y. 1985), aff'd 818 F.2d 145 (2nd Cir. 1987).

39. *In re Paoli Railroad Yard PCB Litigation,* 706 F. Supp. 358 (E.D. Penn. 1988).

40. *Arnett v. Dow Chemical Corporation,* California Superior Court no. 729586, March 21, 1983.

41. Elliott, E.C., Toward incentive-based procedure: three approaches for regulating scientific evidence, presented at Proceedings of the Yale Law School Program in Civil Liability, New Haven, CT, April 8-9, 1988.

42. Austrian, M.L., Significant evidentiary issues concerning expert testimony: a case study of the bendectin litigation, presented at

Defending Toxic Tort Litigation: Seventh Annual Law and Science Defense Seminar, San Francisco, October 27-28, 1988.

43. Gleeson, J.G. and Shelton, H., Reasonable scientific certainty: a proposed evidentiary standard for the disqualification of experts in toxic tort litigation, *DRI Toxic Tort Litigation*, 4, 1, 1986.

44. Black, B., Causation case law - where theory and practice diverge, in *Causation and Financial Compensation for Claims of Personal Injury from Toxic Chemical Exposure: Conference Proceedings*, Novey, L.B., Ed., Georgetown University Institute for Health Policy Analysis, Washington, DC, 1986, 233.

12
DIOXINS AND AGENT ORANGE

The story of dioxin is not merely the story of a chemical. It is the story of public perception and popular fear; of the first 20 years of our environmental regulatory history; of the beginning and evolution of federal policies regarding cancer; of toxicological relationships between animals and people; of environmental activism; of whole towns purchased by the federal government; and even of the aftermath of an unpopular war. Never since the major plagues and infectious diseases have so many policies, vested interests, and emotional commitments been made to a single potential health hazard. Yet, unlike the situation with AIDS, tuberculosis, bubonic plague, or asbestos not a single human being is known to have died or to have suffered a life-threatening illness as a result of dioxin exposure.

The herbicide 2,4,5-T, used in Vietnam as a defoliant and contaminated with small amounts of 2,3,7,8-TCDD (one of many dioxins) was known as Agent Orange, because of the color band that surrounded the drum in which it was packed. By the time the Vietnam veterans' lawsuits against the major manufacturers of the chemical began in the early 1980s, the term Agent Orange was a universal part of the common vernacular. It could never have gotten there without the popularization by newspapers, magazines, and television. It signified death and cancer, specifically brought to those serving in Vietnam by a callous, indifferent military.

The Agent Orange story was a media dream. It presented an ever popular view of the military industrial complex conspiring to produce a highly toxic material, caring little about the health consequences to those in its destructive path. It was also the story of popular indifference about our Vietnam veterans and an opportunity to expiate our national guilt over their treatment. That graphic picture of this chemical and its consequences was the product of media information. Much of that information was simply incorrect or overstated. Cancer rates in veterans were not higher. Even actual exposure to Agent Orange in those claiming its health effects had often not occurred. And, the levels of dioxins to which soldiers could have been exposed were extremely low by comparison with levels of workers manufacturing phenoxyherbicides and those exposed in other mass poisonings. I personally reviewed medical records of two children, alleged victims of Agent Orange, who had deformities of their upper extremities. They had been showcased on television documentaries as living proof that their father's Agent Orange exposure had caused the genetic damage that became their birth defects. A thorough review of their family history told a very different story. These defects had existed for generations and included an uncle and grandfather. They were longstanding inherited disorders, clearly unrelated to the father's Vietnam experience.

During the height of the Agent Orange story, the public's perception was tested in multi-city research polls. Trying to assess the potential outcome of lawsuits against them, the manufacturers of Agent Orange, carried out extensive public opinion surveys. They found that emotions and perceptions were so intense that the public had conclusively made up its mind. Even if solid incontrovertible scientific facts were presented to the jury showing that Agent Orange had not caused the veterans' problems, the jury would not be convinced. This intensity of public perception, notwithstanding the facts, made a fair trial impossible. The companies, as a result, settled the cases for $180 million.

Whether or not they should have paid money to the veterans is not the issue here. What is striking is the discrepancy between the science of dioxin and Agent Orange and the public's perception. Misinformation and partial information had irrevocably distorted public opinion.

"Dioxin" became a word that invoked images of fear and death. The major "poison" in Agent Orange, people linked it irreversibly to the Vietnam veterans and to their claimed residual illnesses. Yet more recently, people have become understandably confused in learning that intensive studies of Vietnam veterans have been unable to confirm what "everyone knew"—that Agent Orange had caused cancers and other illnesses. Thus, popular certitudes about health effects of chemicals generally, and dioxins in particular, are sometimes prematurely developed to be modified later as better research, new information, and more accurate press coverage come about.

Dioxins are among the most discussed chemicals involving our environment today. The public at large has its intense and active concepts of what dioxins are and do. Much of that comes from television or the popular press. Scientists have their understanding, based upon actual research in animals and humans. Regulators have theirs, based upon their interpretation of prudence in the absence of full knowledge and their concepts of what people want. How do we sort all of this out? What do we know about dioxins? What are the scientific facts? How do they depart from the popular perceptions?

STRUCTURE OF DIOXINS

Dioxins are not a single chemical. They are a family of chemicals that contain two benzene rings connected by a pair of oxygen molecules. When chlorine atoms are connected to the benzene rings, we have a chlorinated dioxin. The particular chlorinated dioxin of greatest interest and concern is 2,3,7,8-tetrachlorodibenzo-p-dioxin (generally abbreviated TCDD). For the discussion to follow, the entire family of chlorinated dioxins will be abbreviated PCDD, standing for polychlorinated dibenzodioxins. 2,3,7,8-TCDD will be abbreviated TCDD. The structures of dioxins generally and 2,3,7,8-TCDD are shown in Figure 12.1.

SOURCES OF DIOXINS

The first significant source of dioxin was found to be the manufacturing of an herbicide 2,4,5-trichlorophenol (2,4,5-T). This was the defoliant used in Vietnam, later called Agent Orange. PCDDs come from many places. They are found in products of combustion, both commercially and residentially. One can find PCDDs in chimneys of wood furnaces, in municipal incinerators, in the exhaust of automobiles and in foods. Fish particularly have PCDDs, even those from rivers and lakes with no identifiable industrial source.[1] Because of these many sources, PCDDs are also universally found in us. Americans all have small levels of PCDDs in their blood and fatty tissue—five to seven ppt on average—and we probably have had these levels for many decades.[2] The reason we have only recently become aware of this is because levels in our environment and in our tissues are so small that until relatively recently technology did not permit us to measure them (see Chapter 8).

The question arises then, does the mere fact that these are present pose a threat, or does the fact that their levels are so low provide some measure of comfort? It is this question that engenders extensive debate and profound passions. Yet it does remain a question with no absolutely provable answer. The best one can do is to draw inferences from available scientific studies. When we do that it is my impression, shared by many toxicologists and a growing belief among regulators, that intense concern over extraordinarily minute levels is a misdirected worry.[3,4] At those levels of ppt we find thousands of chemicals in our bodies and environments, many of which at higher levels are toxic or carcinogenic in animals. Furthermore, they have been there for some time despite which our longevity increases and overall cancer rates improve. But the debate

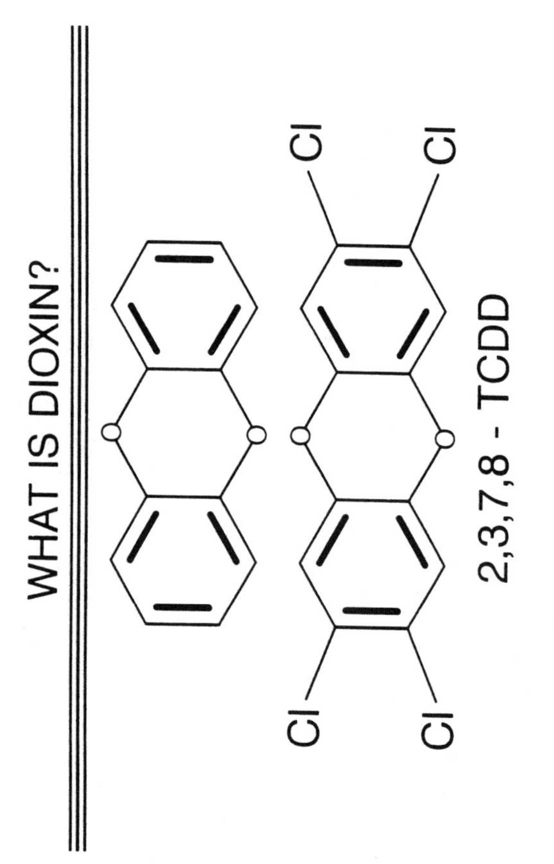

Figure 12.1

goes on with loud voices protesting the smallest and last remnant of dioxin. The sheer emotionality of this issue highlights the personal interests and, at times, irrationality that pervades the discussion. Here we will explore the issue, noting what we do and do not know, and what we can opine or assume based upon current levels of knowledge.

RELATIVE TOXICITY—HUMAN AND ANIMAL

If asked, most people might believe that dioxins are responsible for widespread death among people and surely for death or disease in Vietnam veterans. The reality is that to date, no human is known to have died from exposure to TCDD and the only clearly ascertained chronic effect in man is chloracne—a skin condition resembling a severe case of adolescent acne.[1,4,5] If that is so, then why the intense concern about these chemicals? Concern about dioxins, particularly TCDD, developed, because of their extreme toxicity, both acute and chronic, in certain animals. It was this toxicity that gave them the label "the most toxic substance known to man," a modifier commonly attached to TCDD in popular writing.

While it is true that TCDD is extraordinarily toxic to some animals, its toxicity in humans appears to be far less. Additionally, there exists wide variation among various animal species. When, on July 10, 1976, an explosion occurred at a chemical plant in Seveso, Italy, 1 to 4 pounds of dioxins were spread across a vast area affecting as many as 36,000 people. In the immediate vicinity of the accident animals, large and small, died. Cows, horses, rabbits, sheep, and chickens all died. By contrast, despite identical exposures, no human being died.[6,7] The dose makes the poison (see Chapter 4). Thus, this is an instance in which human sensitivity, at least to dioxin's acute effects, was proven to be less.

In the laboratory, guinea pigs are the most sensitive animals used, so much so that in guinea pigs that label for TCDD "the most toxic substance known" is justified. As little as 0.6 μg/kg kills half of the guinea pigs tested. By contrast, the hamster is 5,000 times less sensitive, requiring 3,000 μg/kg to kill half of the animals. Thus, dioxin manifests significant interspecies variation in toxicity. If even rodents differ so markedly from one another, where do humans fall on that spectrum? Are we more like guinea pigs or like hamsters; or, are we not like either?

In fact, humans seem to be less sensitive to the acute toxic effects than most animals, but exactly how sensitive we are is not known precisely.

The Seveso accident and experience from industrial exposures give some indication of our relative sensitivity, but inexact estimates.

In one direct experiment in people, a group of prisoner volunteers were treated with skin applications of TCDD at concentrations ranging from 200 to 8,000 ng (3-114 ng/kg for a 70 kg person) with a repeat dose 2 weeks later. No one in this group developed any symptoms or findings indicative of dioxin toxicity.[8] In particular, they did not develop chloracne, the characteristic skin condition caused by TCDD. In a second experiment, 107,000 mg/kg were used. Then, 8 of 10 did develop chloracne.[9] It is apparent from this that humans are far less sensitive to this TCDD effect than is the hamster or the rabbit, but how, exactly, human sensitivity compares with that of animals' cannot be determined from these data.

These studies addressed the question of acute toxicity—the production of adverse effects on the body or of death shortly following exposure. As for most environmental contaminants, the concern about TCDD is over chronic effects, particularly cancer and reproductive effects. To assess this issue we can draw from experience taken from industrial exposures and accidental exposures to human populations.

HUMAN EPIDEMIOLOGY

The dioxin epidemiological literature is confusing, inconclusive, and variable. Over 100 human studies have not established the human carcinogenicity of dioxin.[5,10,11] Their wide variability and, at best, weakly positive relationships does suggest that if dioxin is a human carcinogen, it is likely a very weak one.

Richard Monson, in his book *Occupational Epidemiology,* captured the essence of the human studies when he said,

> No body of epidemiologic literature is more confusing than that on dioxin. Seldom have such strong opinions been based on such weak epidemiologic data.[10]

Much of the "dioxin" epidemiological literature suffers from several overriding flaws. The first—confounders—plagues all cancer epidemiology in which an independent variable (potential cause) is, at best, weakly associated with an excess risk. It is impossible to control for all of the confounders that might influence small risks. Dietary fat, dietary fiber, viral causes, vitamins C, E, and A, natural dietary carcinogens (see

Chapter 8), and countless others factors have been linked epidemiologically to cancer risks. Almost never are these controlled for in dioxin studies. A second flaw involves multiple comparisons. We saw in Chapter 5 how subdividing cancers into multiple sites and types inevitably finds associations purely for statistical reasons. Such multiple comparisons commonly produce positive associations by chance alone. By using standard statistical methods, if only one comparison is made, a positive finding has only a 5% chance of being a false positive. By contrast, if ten comparisons are made, that chance rises to 40%. Multiple comparisons, therefore, by their very nature encourage the finding of false positive associations. That is likely what we are seeing in many of these studies.

A third flaw is also found in many environmental epidemiological studies-control of the independent variable—in this case dioxin exposure. In only a few out of more than a hundred dioxin cancer studies has dioxin exposure been established through biological monitoring (blood or fat levels of dioxins). In many cases, the connection between the dioxin cohort and actual dioxin exposure is so tangential as to be utterly speculative and without foundation. Finally, this literature is plagued by a disability peculiar to dioxin and to a few other politically active environmental topics, notably lead and asbestos. An intense effort to prove a bias has crowded the literature with marginal studies designed to justify public policy.

This epidemiological literature can be divided roughly into studies dealing with four types of cohorts: occupational groups having a presumed, but unproven, exposure to dioxins; studies of Vietnam veterans, in which Agent Orange exposure is either presumed or proven through biological monitoring (blood dioxin levels); studies evaluating people after major industrial accidents, most notably that in Seveso, Italy; and studies of pesticide manufacturers with significant dioxin exposure, either presumed or proven through biological monitoring.

The first of these sets of literature is replete with speculative, but unclear exposures. For example, numerous studies have examined the incidence of cancer and other diseases in different occupational groups presuming that pesticide users, foresters, and farmers have occupational exposures to dioxins.[13-20] Results of these studies have varied widely in findings of excess cancers by occupation, type of cancer, and involved organ system. Antidioxin advocates have extracted each positive association from every study and used those collectively to argue for dioxin-related cancers.[21] In fact, these data say little about dioxin since dioxin exposures, either the existence or the extent, are not documented in these studies. And, if these occupations were actually a surrogate for dioxin exposure levels, the shear variability of these data would argue more loudly against dioxin-mediated effects than for them.

The studies of Vietnam veterans have, to date, found no consistent relationships between service in Vietnam and cancers attributable to dioxin exposure.[22-32] Recent studies have measured dioxin blood levels in Vietnam veterans and attempted, unsuccessfully, to correlate those with excess cancers.[26] Thus, despite an intense quest and popular certitude that Vietnam service has produced dioxin-related cancers and death, that has not been proven. Political pressures have, however, ensured that many millions of dollars will be spent for the next decade or more studying this commonly perceived relationship. Because such studies cannot be adequately controlled, they will never establish or refute a causal relationship between Agent Orange and cancers, but it is certain that each new study will bring new associations (for statistical reasons alone) that will be hailed by Veterans groups and their spokespeople as proof of a connection.

The third group of epidemiological studies involve the residents of Seveso, Italy. In the Seveso explosion between 1 and 4 pounds of TCDD was released and spread about the countryside. Severe episodes of chloracne, some requiring children to be hospitalized, were the immediate, acute effects in people. There were, in all, about 200 cases of chloracne in people. People, however, did not die, again suggesting a higher resistance to the most serious acute toxic effects of TCDD by comparison with the animals that were killed.

It was estimated that 750 people were heavily exposed; 4,700 moderately; and 32,000, mildly from this event. The extent of TCDD contamination from this event was, indeed, massive by comparison with the ppt and ppq concerns of today. In Seveso soil levels of TCDD ranged, in the most contaminated area, from 400 to 20,000 mg/acre, hundreds to thousands of times higher than potential levels associated with Agent Orange in Vietnam and hundreds of thousands of times higher than the lower EPA water limits. Even the cows' milk in the high contamination areas of Seveso contained 1 to 7 ppb of TCDD.

Dioxin levels in the blood of the Seveso residents have now been assessed. Levels are quite high, ranging, in the highest exposure group, from 828 to an astounding 56,000 ppt, the highest human dioxin level ever seen. As Tollefson correctly points out, "the Seveso residents clearly have some of the highest TCDD exposures known to date."[8] Thus, exposures have been proven and shown to be quite high. Despite these relatively high exposures, the long-term and chronic toxicity evaluations in Seveso have been quite encouraging. Women pregnant at the time of the accident did not have an unusually high incidence of either miscarriages or of children with birth defects. And, though it is still too early to know for sure, to date, there is no significant increase in overall cancer mortality in the Seveso residents.[7]

Recently, three long-term studies of heavily exposed herbicide manufacturing workers were reported. In one of these studies actual dioxin blood levels were measured.[33] In another, they were measured, not in the study population, but in a sampled group of coworkers with seemingly identical occupational exposures.[34] In the third, the fact of a heavy exposure (similar to the Seveso event) resulted from a 1953 industrial accident followed by monitoring of blood dioxin levels.[35]

The study by Fingerhut and colleagues published in the *New England Journal of Medicine*, reviewed approximately 5,172 workers, exposed to TCDD for varying lengths of time, who were followed for up to 30 years and had their cancer mortality assessed. There were excess risks of certain cancers, primarily respiratory. Certain groups had somewhat higher cancers rates, but rates far lower than those found in some other studies. Other groups had no higher rates or even lower rates.[33] While this study correlated certain cancers with dioxin exposures, design difficulties and findings limit the ability to ascribe causation. First, there was no dose-response relationship as one might expect if causation were clearly present. Second, exposures were not to dioxins alone but to a wide range of occupational chemicals. Thirdly, data from multiple work sites were combined despite potentially great differences among those sites. Finally, controls for confounders, such as smoking, were limited.

In a related paper, Manz and colleagues studied 1,583 workers employed in a German herbicide plant. Comparing the workers to either gas workers or to the West German population-at-large, there were small, statistically significant excesses in cancers among the herbicide workers. The sites varied and differed from those in the Fingerhut study, limiting our ability to point to a common cause. Also, confounders such as benzene exposure were noted, but uncontrolled.[34]

Finally, the Zober study found no excess risk of cancer among dioxin-exposed workers. When workers who had developed chloracne were compared with controls, there was still no excess risk. But, when this group was further fractionated into those exposed twenty or more years earlier, there was a small excess.[35]

These new studies of industrial workers have, for the first time, attempted to correlate, not only presumed exposures, but either biologically proven dioxin exposures or strongly inferred exposures and a variety of malignancies. Individuals in all three of these studies were likely heavily exposed. In the study by Fingerhut and colleagues, average dioxin blood levels exceeded 100 ppt and the groups in which excess cancer were identified had levels of 1,000 ppt.[33] In the study by Manz and colleagues actual dioxin levels in the cohort were not measured, but levels were measured in a group of workers with presumed identical exposures. The heavily exposed group had average dioxin levels of 296 ppt.[34]

These studies found some correlations between the heaviest exposures and the longest latency periods and certain excess cancers. They leave us with three new studies suggesting a small increased risk, but by no means proving it.

Editorial comment in the *New England Journal of Medicine* following the Fingerhut article predicted that this article would be cited by scientists on both sides of the dioxin debate. Those who would lessen the regulations would claim that this proved that dioxin was not a human carcinogen. Those who crusade against dioxin would use the same data to strengthen their argument. That prediction has come true. Such diametrically opposite interpretations of the same data could not have happened, of course, if the data were interpretable in an unequivocal way. They were not.

Design difficulties and contradictory findings limit our ability to formulate clear answers from these studies. What they do show, however, is that if dioxin is a carcinogen, it is weakly so and seems to require significant and prolonged exposure to produce cancer.

So, if TCDD is a human carcinogen, it is not likely a very potent one. Moreover, the studies performed have involved heavy industrial and accidental exposures, vastly greater than exposures associated with background environmental contamination.

Nonscientists sometimes wonder: why can't we just find out? Why are there all these disagreements among scientists? The reasons for these variations in findings are based less upon the points of view of scientists than upon problems inherent in doing this research. Some answers to this were discussed in Chapter 5. First of all, it is quite clear from these studies that TCDD is not a potent human carcinogen, or, at least, it is vastly less so in people than in the most sensitive animals. If that were not so, the relationship between TCDD and human cancers would, by now, no longer be a matter of debate, it would be clear. When we are dealing with, at best, a weak carcinogen; its effects are not easily picked up over the background cancer rate. Simply put, one in four people will get cancer: that is the rate at which it generally occurs. If an agent increases that rate by 0.0001, then the rate goes from 0.25 to 0.2501, a statistically small number, only visible if thousands of people are studied. Furthermore, the numbers of potential factors, chemicals among them, in our air, food, water, and our workplaces, that may have that kind of small impact, number in the thousands. It becomes nearly impossible to control for those, to separate and sort out the single factor of concern—in this case TCDD. Finally, we often have limited data about actual TCDD exposures and are left with estimating those—a process bound to produce scientific uncertainties.

MARGINS OF SAFETY

It has been estimated that workers who manufacture, apply or formulate the herbicide 2,4,5-T have, since its introduction in the 1950s, taken into their bodies 4 μg per day.[6] Most studies of those workers have not identified any clear excess risk of cancer or other serious long-term disease, although the Fingerhut, Mantz, and Zober studies noted above suggest a weak relationship. For comparison purposes, the EPA has determined as acceptable for ingestion, levels of 0.002 μg/day, approximately 2,000 times less than those industrial exposures. It seems likely that there are very large margins of safety built into the these regulatory guidelines.

Another comparison is useful. The dioxin blood level found by Fingerhut to be associated with an excess cancer risk is 3,000 ppt. An EPA guideline raises concerns about dioxin levels in fish when those levels exceed 7 ppt. In a study by Svenson, Swedish fishermen were found to have blood levels of 7 ppt (the customary background level in the United States) when they ate 3 pounds of fish per week containing dioxin ranging from 8 to 18 ppt.[36]

The dioxin debate marches inexorably into the 21st century. Though it is unlikely that any more epidemiological studies will bring us any closer to "true" answers than we are today, such studies will no doubt continue. The best hope for putting to rest, once and for all, the question of dioxin carcinogenicity, lies in basic research. Ultimately, exploration and resolution of fundamental mechanisms of carcinogenicity promise to solve the dilemma and end the debate.

REFERENCES

1. Agency for Toxic Substances and Disease Registry, *Toxicological Profile for 2,3,7,8—Tetrachlorodibenzo-p-dioxin,* National Technical Information Service, Springfield, VA, 1989.

2. U.S. Environmental Protection Agency, Office of Toxic Substances, *Broad Scan Analysis of the FY82 National Human Adipose Tissue Survey Specimens. Volume I. Executive Summary,* U.S. Environmental Protection Agency, Washington, DC, 1986.

3. Kimbrough, R.D., How toxic is 2,3,7,8—tetrachlorodibenzo-dioxin to humans? *J. Toxicol. Environ. Health*, 30, 261, 1990.

4. Houk, V.N., Dioxin Risk Assessment for Human Health: Scientifically Defensible or Fantasy?, presented at the Twenty-Fifth Annual Conference on Trace Substances in Environmental Heath, University of Missouri, Columbia, MO, May 21, 1991.

5. Lilienfield, D.E. and Gallo, M.A., 2,4—D, 2,4,5—T, and 2,3,7,8—TCDD an overview, *Epidemiol. Rev.*, 11, 28, 1989.

6. Ottoboni, M.A., *The Dose Makes the Poison,* Vinciente Books, Berkeley, CA, 1984.

7. Bertazzi, P.A., Zocchetti, C., Pesatori, A.C., Guercilena, S., Sanaioco, M. and Radice, L., Ten-year mortality study of the population involved in the Seveso incident in 1976, *Am. J. Epidemiol.*, 129, 1187, 1989.

8. Tollefson, L., Use of epidemiology data to assess the cancer risk of 2,3,7,8—trichlorodibenzene-*p*-dioxin, *Regul. Toxicol. Pharmacol.*, 13, 2, 1991.

9. Bond, A.G., Bodner, K.M. and Cook, R.R., Phenoxy herbicides and cancer: Insufficient epidemiologic evidence for a causal relationship, *Funda. Appl. Toxicol.*, 12, 172, 1989.

10. Monson, R.R., *Occupational Epidemiology*, 2nd Edition, CRC Press, Boca Raton, FL, 1990.

11. Corrao, G., Calleri, M., Carle, F., Russo, R., Bosia, S. and Piccioni, P., Cancer risk in a cohort of licensed pesticide users, *Scand. J. Work Environ. Health,* 15, 203, 1989.

12. Eriksson, M., Hardell, L. and Asdami, H.O., Exposure to dioxins as a risk factor for soft tissue sarcoma: a population-based case-control study, *J. Natl. Cancer Inst.,* 82, 86, 1990.

13. Weisenburger, D.D., Environmental epidemiology of non-Hodgkin's lymphoma in eastern Nebraska, *Am. J. Ind. Med.,* 18, 303, 1990.

14. Green, L.M., A cohort mortality study of forestry workers exposed to phenoxy acid herbicides, *Br. J. Ind. Med.,* 48, 234, 1991.

15. Pearce, N.E., Smith, A.H. and Fisher, D.O., Malignant lymphoma and multiple myeloma linked with agricultural occupations in New Zealand cancer registry-based study, *Am. J. Epidemiol.,* 121, 225, 1985.

16. Persson. B., Dahlander, A., Fredriksson, M., Brage, H.N., Ohlsin, C.G. and Axelson, O., Malignant lymphomas and occupational exposures, *Br. J. Ind. Med.,* 46, 516, 1989.

17. Wingren, G., Fredrikson, M., Brage, H.N., Nordenskjold, B. and Axelson, O., Soft tissue sarcoma and occupational exposures, *Cancer,* 66, 806m, 1990.

18. Hardell, L. and Eriksson, M., The association between soft tissue sarcomas and exposure to phenoxyacetic acids, *Cancer,* 62, 652, 1988.

19. Alavanja, M.C., Blair, A., Merkle, S., Teske, J., Eaton, B. and Reed, B., Mortality among forest and soil conservationists, *Arch. Environ. Health,* 44, 94, 1989.

20. Wigle, D.T., Semenciw, R.M., Wilkens, K., Riedel, D., Ritter, L., Morrison, H.T. and Mao, Y., Mortality study of Canadian male farm operators: non-Hodgkin's lymphoma mortality and agricultural practices in Saskatchewan, *J. Natl. Cancer Inst.,* 82, 575, 1990.

21. Affidavit of Cate Jenkins, Ph.D., in Cv-89-03361 (E.D.N.Y.) (JBW), [B-89-00559-CA (E.D.TEX.0], September 1991.

22. Dalager, N.A., Kang, H.K., Burt, V.L. and Weatherbee, L., Non-Hodgkin's lymphoma among Vietnam veterans, *J. Occup. Med.*, 33, 774, 1991.

23. Breslin, P., Kang, H.K., Lee, Y., Burt, V., Shepard, B.M., Proportionate mortality study of U.S. Army and U.S. Marine Corps veterans of the Vietnam War., *J. Occup. Med.*, 20, 412, 1988.

24. Wolfe, W.H., Michalek, J.E., Miner, J.C., *An Epidemiologic Investigation of Health Effects in Air Force Personnel Following Exposure to Herbicides: Summary Mortality Update,* Brooks Air Force Base, TX, 1989.

25. Centers for Disease Control Vietnam Experience Study, Health status of Vietnam veterans, II, Physical health, *JAMA,* 259, 2708, 1988.

26. Roegner, R.H., Grubbs, W.D., Lustik, M.B., Brockman, A.S., Henderson, S.C., Williams, D.E., Wolfe, W.H., Michaelk, J.E., and Miner, J.C., *An Epidemiologic Investigation of Health Effects in Air Force Personnel Following Exposure to Herbicides, Air Force Health Study,* Brooks Air Force Base, TX, 1991.

27. Watanabe, K.K., Kang, H.K. and Thomas, T.L., Mortality among Vietnam veterans: with methodological considerations, *J. Occup. Med.*, 33, 780, 1991.

28. Kang, H., Enziger, F., Breslin, P., Feil, M., Lee, Y. and Shepard, B., Soft tissue sarcoma and military service in Vietnam: A case study, *J. Natl. Cancer Inst.*, 79, 693, 1987.

29. Clapp, R.W., Cupples, L.A., Colton, T., and Ozonoff, D.M., Cancer surveillance of veterans in Massachusetts, USA, 1982-1988, *Int. J. Epidemiol.*, 20, 7, 1991.

30. Thomas, T.L. and Kang, H.K., Mortality and morbidity among army chemical corps Vietnam veterans: a preliminary report, *Am. J. Ind. Med.*, 18, 665, 1990.

31. Kogan, M.D. and Clapp, R-W., Soft tissue sarcoma mortality among Vietnam veterans in Massachusetts, 1972 to 1983, *Int. J. Epidemiol.*, 17, 39, 1988.

32. Fett, M.J., Nairn. J.R., Cobbin, D.M. and Adena, M.A., Mortality among Australian conscripts of the Vietnam conflict era, II, Causes of death, *Am. J. Epidemiol.*, 125, 878, 1987.

33. Fingerhut, M.A., Halperin, W.E., Marlow, D.A., Piacitelli, L.A., Honchar, P.A., Sweeney, M.H., Greife, A.L., Dill, P.A., Steenland, A.B. and Suruda, A.J., Cancer mortality in workers exposed to 2,3,7,8-trichlorodibenzodioxin, *N. Engl. J. Med.*, 324, 212, 1991.

34. Manz, A., Berger, J., Swyer, J.H., Flesch-Janys, D., Nagel, S. and Waltsgott, H., Cancer mortality among workers in chemical plant contaminated with dioxin, *Lancet,* 338, 959, 1991.

35. Zober, A., Messerer, P. and Huber, P., Thirty-four-year mortality follow-up BASF employees exposed to 2,3,7,8—TCDD after the 1953 accident, *Occup. Environ. Health*, 62, 139, 1990.

36. Svenson, B.G., Nilsson, R.N., Hansson, M., Rappe, C, Akesson, B. Skerfving, S., Exposure to dioxins and dibenzofurans through the consumption of fish, *New Engl. J. Med.*, 324, 7, 1991.

13
"SICK BUILDINGS"

CASE STUDIES

Fifteen employees of an accounting firm in Los Angeles recently sued the developer of their office building. They had all developed a variety of physical complaints after caulking, which contained a number of organic chemicals, was applied to the air conditioning ducts. They contended that the complaints persisted even after the caulking was removed, the building aired, and the chemicals were gone. The case ultimately settled for in excess of a million dollars.

In another suit, a number of teachers in a Florida school claimed that formaldehyde from new carpeting produced both acute and chronic illnesses. One was awarded permanent workers' compensation benefits.

In a high school in northern Virginia, there were at least 100 episodes of students fainting during the school year 1990-91. Additionally, students complained of a wide range of symptoms, most manifest while they were in the school and particularly when they were in several newly enclosed rooms. Enraged parents have picketed, protested, and sued demanding that the school be closed, certain that their children are being poisoned. Comprehensive evaluations thus far have found only that the rooms are somewhat stuffy. Physically, the students have no identifiable building-related dysfunction, and the air in the rooms is not laden with chemicals.

A school district in central Pennsylvania underwent asbestos abatement in the summer of 1990. Part of the process involved removal of floor tiles for which the contractor used large amounts of a petroleum-based organic solvent. Following that and throughout the school year teachers and children complained of a variety of symptoms: headaches, eye and throat irritation, frequent colds, cough, and others. Multiple evaluations of the school by engineering consultants found small amounts of residual chemicals, insufficient, they opined, to produce illness. Finally, concerns peaked and tempers flared to such a point that the school was closed. Further investigation showed that the solvent had penetrated deeply into the cement floor. While the amounts were insufficient to cause serious health effects, they were sufficient to produce some irritation in sensitive individuals and a noxious smell, a factor that can itself produce symptoms.

Finally, even the EPA building itself in downtown Washington, D.C., has been the subject of attacks upon its indoor air quality. In 1980 complaints began to develop after new carpeting was installed. By 1991 a class action suit on behalf of 5,000 employees had been filed, alleging that chemicals in the carpeting, particularly 4-phenyl cyclohexane, had caused a wide variety of physical complaints and physical disorders. Support for this allegation came predominantly from a series of subjective questionnaires filled out by employees and the intense, nearly messianic zeal and certitude of two employees who were certain that this chemical was the culprit.

These are a few of the "sick building" matters in which I have been consulted. They illustrate the ranges of potential causes from chemicals to rumors that underlie these complaints.

BACKGROUND: PERCEPTION OF THE SICK BUILDING SYNDROME

The term "sick building" syndrome is now part of the common vernacular, but it is misleading, for it suggests a common building-related cause for any group of workers or students who develop symptoms. The term "sick building" is put in quotation marks because its medical and industrial hygiene meaning is highly uncertain, while its popular definition seems well-understood and commonly accepted.

The public is convinced that indoor air contains substances, particularly chemicals, that produce a variety of symptoms or ailments. Sometimes that is so (two of the five cases above). But studies have shown that only rarely can such physical complaints be linked to indoor air problems. Tightly sealed buildings, the result of our 1970s energy policy, or contaminants such as molds and, rarely, chemicals have occasionally been proven to cause discomfort among workers. Far more common are the indoor air complaints without a provable cause which are rising rapidly. Dr. Jan A. J. Stolwijk, Department of Epidemiology and Public Health, Yale University School of Medicine, a student of this problem, recently noted that between 500 and 5,000 buildings have been evaluated for these complaints, but that "relatively few of these investigations have led to clear conclusions which could be implemented into effective corrective action."[1]

It is undoubtedly true that public awareness of these "problems" with indoor air will intensify complaints and distress. Odors formerly ignored but now perceived as hazardous, will generate intense demands for corrective measures. The difficulty lies in the very definition of "the problem." Clearly, if people perceive a problem, that in itself can be a problem. But corrective approaches require a more specific understanding of air quality issues and an ability to separate those factors that create a true health hazard from those that produce annoyance. Here risk communication will also play a critical role. Odors are ubiquitous and people respond quite differently to them. Once investigation and reasonable remediation is completed, some building residents may still notice odors that are toxicologically irrelevant, but not remediable. Explaining that to individuals concerned about chemicals is a challenging task of growing importance. What are the scientifically established facts about indoor air and its effects upon health? How do we separate factors that are perceived as a hazard from those that actually are?

Until quite recently the term "syndrome" was never part of the popular vocabulary. Now we have "toxic shock syndrome," "premenstrual syndrome," "chronic fatigue syndrome," "20th-century syndrome," "chronic yeast syndrome," "sick building syndrome" and many others. Americans' intense health awareness has publicized these "syndromes," some of which are medically established many are not. Every collection of physical complaints, it seems, becomes a new syndrome. The very use of the term "syndrome" conveys a lay meaning suggesting commonality of complaints and cause. To be sure, certain syndromes are well known in medicine and causally understood. Today, however, popular perceptions and media discussions find new syndromes at an ever-increasing pace. Popular understanding of these new syndromes frequently precedes a clear medical understanding of any common characteristics or cause. Not infrequently, perceptions are wrong; no commonality exists. Thus, the popularization of "syndromes" can be a reflection of the discrepancy between common perceptions and scientific realities.

The public believes that buildings can make people sick, and that they often do. The *Washington Post,* September 17, 1989, claimed

> Environment: Your Office May Be Hazardous: More and more Americans are becoming aware of air-quality hazards in the work place, and terms such as "sick building syndrome" and "Legionnaire's disease" have entered the vernacular. But air pollution is only one threat among the many in the modern office. Recent research has revealed numerous other environmental factors with direct and indirect effects on human health and performance.

The *Los Angeles Times* wrote on January 14, 1991,

> Employee Group Seeks Cure for "Sick" County Courthouse. Environment: Death and illnesses of some courthouse workers have raised concerns about asbestos and air-quality problems at the building. But the county denies it is a hazard. The deaths of several courthouse workers and judges, and the mysterious illnesses that have plagued other employees, have sent Superior Court Clerk, Chuck Martin looking for an antidote to the "sick building syndrome."

And, from the *San Francisco Chronicle,* November 23, 1990,

> Sick Building Syndrome A Growing Risk to Workers: Jennifer, who sells specialized services for a New England-based computer manufacturer, used to be an active young adult before she moved into her new company's office complex more than a year ago.

A number of scientists also believe that indoor air problems pose a significant health threat. In a sample of office workers conducted by Honeywell Technalysis, 24% of 600 workers perceived air quality problems in their offices.[2] Nearly 10% perceived the problems to be very serious, and 20% perceived them to be interfering with their productivity. Woods took these figures and extrapolated them to the nation. He suggested that 800,000 to 1,200,000 commercial buildings in the United States have problems that lead to the "sick building" syndrome and affect 30 million to 70 million exposed occupants.[3] The problem here is that the cause and effect is unclear. No doubt the survey is accurate. People perceive an indoor air problem. Woods then assumes that it is, in fact, the indoor air that is causing the perceived problem and moves from there to broad extrapolations and expansive recommendations for correction. It is just as likely that people's perceived cause of their symptoms is only occasionally the true cause. It is just as likely that emotional factors (such as job dissatisfaction) combined with a heightened awareness of indoor air issues (reporting biases) leads to responses indicting indoor air quality. We discussed in Chapter 5 the problems inherent in symptom questionnaires.

The perception that buildings are dangerous may have begun with the deaths at the Bellevue Stratford Hotel in Philadelphia, Pennsylvania, in 1976. There, 182 persons attending an American Legion convention became ill and 29 died of a mysterious lung ailment—a form of pneumonia. Careful sleuthing by the Bureau of Epidemiology of the

Centers for Disease Control ultimately identified the culprit as *Legionella pneumophila* bacteria residing in the air conditioning system of the building and spewed forth as a microbial aerosol whenever the unit was on.[4-5] Here, a clear culprit was identified, a specific disease found. This building at that time was indeed hazardous.

The next major popularly publicized indoor air issue was that associated with urea-formaldehyde foam insulation. In the mid-1970s, when energy conservation was of paramount concern, urea-formaldehyde foam insulation became the rage. It was sprayed into walls of homes throughout the country until the complaints and lawsuits began. People described a variety of irritant effects and some very serious illnesses. Many were associated with the foam insulation, many with formaldehyde from insulation in mobile homes. How many of these symptoms were actually due to chemicals from the foam is unclear. As is generally the case, the line between toxin-induced complaints and alternate causes, including fear and lawsuits, eventually became blurred as publicity about this perceived hazard increased. Urea-formaldehyde foam insulation was banned by the Consumer Product Safety Commission (CPSC) in 1982 (a decision that was overturned for lack of scientific support). What remained was the popular notion that noxious chemicals in the air of energy-efficient, well-insulated homes threatened our health.

The current media coverage of these building-related health issues reflects the general confusion surrounding the matter. They may mix deaths, asbestos, radon, illnesses, and symptoms in a single story leaving the impression that they are somehow all linked by a common disorder of indoor air. Most commonly, stories about indoor air are covered by local papers in response to complaints by workers or families of students. Frequently, there is little scientific input in the story, often because the "problem" is unidentifiable. This leaves the distinct feeling of a mysterious and silent threat, one that is so pernicious that it defies detection.

When people died from contaminated indoor air as did the victims of Legionnaire's disease, the clinical endpoint was unequivocal—death. Most indoor air phenomena are far less well-defined. They involve nonspecific complaints such as tiredness, headaches, difficulty concentrating, dry eyes and mouth—complaints associated with hundreds of causes ranging from run-of-the-mill allergies such as hay fever to life stresses, personality characteristics, job dissatisfaction, noise, lighting, and others. This sheer range of potential causes, confounders, and covariates (job satisfaction or dissatisfaction is obviously a major one in workplace-related complaints) impedes ready investigation. As a result, studies may be biased by the investigator's assurance that indoor air is or is not actually the culprit. Several articles are illustrative. In one, chemicals in indoor air were correlated with symptoms in workers. The investigators concluded that indoor air quality specifically involving environmental causes such as lighting and VOCs affected inhabitants.[6] In another similar study, the investigators added another factor to the evaluation: job satisfaction. They found that it was job satisfaction, rather than indoor air quality that affected symptomatology.[7] These studies illustrate the profound difficulties involved in the study of nonspecific symptoms, the kind associated with most indoor air disorders. They also illustrate the depth of uncertainty and the breadth of misperceptions pervasive even among the medical and scientific communities.

SCIENCE: ILLNESSES ASSOCIATED WITH INDOOR AIR PROBLEMS

The study of indoor air-related disorders is sufficiently new and heterogeneous that the terminology is unclear. "Sick buildings," "tight buildings," and "building-related disease" are used interchangeably with little definitional clarity. The scientific literature suggests that several building-associated conditions should be placed into separate categories according to established causes. These include building-related diseases; tight building syndrome; and building-associated symptoms.

Building-Related Diseases

Building-related diseases are disorders, ranging from mild to severe, that are due to specific, identifiable contaminants of indoor air. Legionnaires' disease is an example. Here a specific bacteria was causal. Certain other organisms, commonly fungi, molds, and thermophilic bacteria contaminating heating and air conditioning systems produce varieties of complaints and disorders. These are generally mild hay fever types of allergies, but they may be more serious, such as asthma or hypersensitivity pneumonia. Other common infectious diseases like colds and influenza may be spread by ventilation systems. When a large fraction of the work force becomes ill from such infections, they, too, can be considered building-related diseases, although it may be difficult to find a specific contributor in the building environment itself. In a study of army recruits living in "leaky" barracks versus those living in "tight," more energy-efficient barracks, it was found that the latter group had a higher frequency of colds.[8] It is certainly reasonable to assume that confined spaces with poor outside ventilation would be better transmitters of respiratory viruses. How and whether that translates to more open and far larger office buildings is, however, not established. To be classified as a building-related disease, there must be clear and convincing evidence that something in the building is causal and, preferably, the agent should be known.

Moreover, the disease or endpoint of this disorder is generally quite clear-cut, rather than a set of nonspecific complaints. It may be death or serious respiratory infection, as was the case in Legionnaires' disease. It may be an epidemic of influenza passed through a work force. It may be an occupational asthma proven by immunological studies of the patient correlated with identification of the causative biological factors found to be originating in the building's ventilation system.

"Sick Building" Syndrome

The "sick building" syndrome implies that a significant fraction of building occupants complain of a variety of building-associated symptoms such as eye and mucous membrane irritation, headaches, fatigue, and sinus congestion. Furthermore, it requires a substantial attack rate: involvement of 20% or more of building occupants, a temporal relationship to the building and improvement with specific corrective measures. When the additional label "tight building syndrome" is added, a causal attribution is also added. This label implies that a problem with indoor air, generally related to poor air-exchange in an energy-efficient building, has been identified. "Sick building" syndrome differs from building related disease in that a specific agent such as bacteria or mold is rarely found. Rather, the carbon dioxide levels may be elevated, indicating poor air circulation and/or exchange, and a variety of indoor air contaminants may be similarly elevated.

Today in our body and health-conscious society physical sensations and symptoms are closely monitored; thus, stuffy air produces symptoms that may become the "tight building syndrome." It has been argued, and likely the arguments have some merit, that our energy-efficient buildings constructed after the early 1970s have sealed the internal environments, permitting a variety of contaminants to linger and accumulate when formerly they would have migrated to the outdoors. While it is clearly true that modern building are more tightly prevented from commingling with outdoor air, it is not clearly true that indoor air today is worse than it used to be. For example, in this country there are vastly fewer smokers now than there were in 1965. Then, conference rooms of office buildings were filled with cigarette and even cigar smoke (hence the expression, "smoke-filled rooms"). What could be more disturbing than a crowded room filled with the irritating chemicals emitted by tobacco smoke? Wasn't that indoor air environment more contaminated than today's, in which ppb of unseen and often unsmelled chemicals are the focus of such intense concern?

Because "tight building syndrome" is associated with nonspecific symptoms and is dependent upon subjective individual questionnaires for its identification, its cause—air contamination or psychological—cannot readily be separated. Others, including Colligan, have commented upon this.[9] Moreover, as reporting of "indoor air problems" intensifies, so will psychological influences and reporting biases. The only way to approach some semblance of true scientific investigation would be through controlled, blinded studies in which the air constituents were varied unbeknownst to building occupants whose symptomatology would be reassessed.

Whether today's "sick" or "tight" building syndromes truly represent defined syndromes is, for all of these reasons, not fully established. The prevalence of complaints alone does not prove that the cause is poor air quality. It could be due to a high pollen count outdoors, a common viral illness, a dissatisfied work force, reporting bias, or other factors. Only an intense, scientifically controlled investigation might distinguish among these alternatives. Moreover, the symptoms do not establish the cause. They typically vary in nature from person to person and they are sufficiently nonspecific (having many possible causes) to render uncertain a common causal attribution. Nonetheless, it is believed that such a syndrome does exist, but there is less consensus about how often and how to prove it, than there is in the case of building-related diseases.

The next situation, building-associated symptoms, is an even softer phenomenon scientifically.

Building-Associated Symptoms

This is the softest group of building-related conditions or complaints. Here, occupants of a building complain of a constellation of symptoms. Tight building syndromes and many cases of building-related diseases may start this way—with complaints. Then, once the cause is found or a causal relationship to the building is identified, it moves into one of the above categories. Most situations involving building-associated symptoms, however, never leave this category. Intensive investigation is unable to elucidate a specific common cause, or the possible cause is too speculative to provide any level of confidence.

CAUSES

Because we are dealing here with an eclectic group of symptoms and disorders, causes are multifaceted. They range from purely emotional factors to infectious viruses and bacteria. Clearly, as perceptions of "bad indoor air" increase, emotional factors grow in importance and it becomes increasingly difficult to sort out the real culprits and to separate symptoms due to perceptions from those due to bona fide contaminants. Below is a brief discussion of some specific ambient factors in indoor air that may affect levels of comfort and could contribute to indoor air symptoms. We have already discussed factors, such as bacteria, that may cause serious diseases and have considered psychological factors that may cause or intensify complaints.

Ventilation and Related Factors

Poor ventilation and/or the type of building ventilation is the factor most often studied when researchers look for a cause of "sick building" syndrome. But when we read the studies we see much doubt and little conclusive evidence about whether ventilation alone or ventilation combined with other factors is a correctable source of the problem. NIOSH notes that in 52% of cases of buildings investigated for indoor air complaints, there were problems with the ventilation.[10] This does not mean, of course, that they were causal.

One of the more frequently cited epidemiological studies by Finnegan and colleagues concludes that there are consistently more symptoms in air conditioned buildings as compared with naturally ventilated buildings. Beyond this, no specific cause, such as the use of humidifiers or the presence of formaldehyde or other chemicals, could be identified.[11] A study by Burge concluded specifically that buildings with ventilation from local or central induction fan coil units had more symptoms than buildings with "all air" ventilation systems; both systems had more symptoms than naturally or mechanically ventilated buildings. According to this study, microbiological contamination from chillers, duct work, or humidifiers (secondary to the ventilation system) resulted in some of the worst symptoms, probably by an allergic or endotoxin-related mechanism.[12] Another investigator, Michael Hodgson, measured a number of environmental characteristics including thermal parameters (dry-bulb temperature, relative humidity, air speed, and radiant temperature),

VOCs, respirable suspended particulates, lighting and noise intensity, and carbon monoxide and carbon dioxide levels. Hodgson found that certain specific "causes" stood out, and these turned out to be lighting and VOCs; layers of clothing and crowding were also related to increased symptoms.[6] In fact, this study could not identify causes, merely correlations (see Chapter 5).

The much-quoted Danish study of 4369 workers in 14 town halls, reported by Skov and associates found no single, clear correlation with "sick building" syndrome symptoms. Temperature and humidity, carbon dioxide and formaldehyde levels, static electricity, dust, microorganisms, VOCs and lighting were among the variables studied. Interestingly, the results did not corroborate earlier findings by Robertson and Finnegan of a higher symptom prevalence in mechanically ventilated buildings than in naturally ventilated ones.[13]

Other researchers in this country do conclude that even if a single etiology is unknown or there are multiple confounding variables, the type of ventilation or poor ventilation has more bearing on indoor air quality than any other factor. Unfortunately, the research is epidemiologically imprecise.

If building ventilation leads to symptoms, the mechanism is unknown; it may be that reduced ventilation may directly affect changes in comfort or it may cause the build-up of chemical pollutants.[14] It has also been suggested that factors other than ventilation, such as poor management, boring work, noise, poor lighting, and temperature may lower the threshold for tolerating or complaining about symptoms.[15]

Odors

Odors, particularly unfamiliar smells, smells viewed as noxious or "chemical" smells, invoke a wide range of emotional responses. One has only to think of responses to the smell of dead flesh. People faint, vomit, develop palpitations, and innumerable other symptoms with such exposures. Is that because the putrefying flesh gives off toxic chemicals? Of course not! It's because it arouses a psychological effect. Today chemical smells arouse psychological effects, because they are perceived to represent a hazard. Whether they actually do or not is a highly chemical-specific matter. Many completely odorless chemicals, e.g., carbon monoxide, can be quite toxic, while other highly malodorous ones (mercaptans) are only minimally so. In building-related symptom complexes, however, odors can be extremely important and promise to grow in importance.

Emotional Causes and Mass Hysteria

We have already discussed the role that psychological factors can play. Specifically, there may be psychosocial issues outside of the workplace and stresses within the workplace that can result in some of the symptoms such as headache or lethargy. "Sick building" syndrome in and of itself is an emotionally charged issue.

There have also been many reported incidents of epidemic anxiety and mass hysteria. These incidents sometimes begin with a remediable indoor air problem, and other times they can be traced not to a building problem but to "problem" individuals. In one acute epidemic, several hundred employees in a state office building in Missouri complained of headache, mucosal irritation, fatigue, odd taste, and dizziness. Extensive investigation revealed no toxic substances or direct cause of the illness. One interesting finding was that the employees who complained of illness were more likely to have perceived unusual odors and inadequate air flow. In any event, the conclusion was that a state of epidemic anxiety was triggered by negative factors in the environment, including poor air quality (crowding, blocked vents, smoking, high temperatures). Reports of illness from coworkers, arrival of emergency vehicles, and evacuation of the building probably led to the escalation of the event.[16]

In another, similar incident employees in a telephone operators' building reacted to what they perceived was a strange odor (or reports of a strange odor) with symptoms of headache, nausea, throat irritation, and even respiratory distress. The incident was dragged out over a period of a month, with evacuations and inspections by California OSHA, the local fire department, and the county hazardous materials management team. When no evidence of toxic fumes, gases, or chemical leaks or spills could be found, and all of the people taken to hospitals were found to be healthy with normal laboratory results, the investigation turned to one individual who, in fact, had spread the epidemic as he moved from one part of the building to another reporting of noxious odors that he interpreted as petroleum distillate poisoning. Here, too, the hysteria escalated with the arrival of emergency vehicles, as workers witnessed fellow employees in respiratory distress.[17]

I, too, have been involved as a consultant in dozens of matters in which an anxiety to perceived but unproven indoor air contamination produced widespread symptoms. It has been pointed out that with the increased awareness of chemicals in the environment and the media attention to indoor air issues, there is a greater tendency for psychological factors to aggravate and even cause outbreaks of illness in the workplace. Often, investigations do not reveal a specific causal agent, even with careful monitoring of air contaminants. But because symptoms often disappear with improvements in ventilation and/or when individuals leave the building, it is difficult to know whether there is a physiological basis for the illness or whether it is a case of mass hysteria or epidemic psychogenic illness. We need to work on developing scientific criteria for distinguishing between illness arising from psychological factors and symptoms resulting from exposure to indoor air pollution or toxic substances.[14]

Products of Combustion, Including Tobacco Smoke

Sometimes combustion products are implicated in building-related illnesses, particularly in cases of respiratory effects. These products consist primarily of carbon monoxide (CO), nitrogen dioxide (NO_2), and sulfur dioxide (SO_2). Sources of these compounds are tobacco smoking, gas ranges, pilot lights, unvented kerosene space heaters, wood and coal stoves, fireplaces, and vehicle emission exhaust. The "contaminant" that probably has received the most publicity is environmental tobacco smoke (ETS), also called passive tobacco smoke (PTS), which refers to the tobacco smoke inhaled by nonsmokers. One should keep in mind that the exposures from passive smoke are mainly sidestream. Though sidestream smoke may have some toxic substances as does mainstream (active) smoke, it is diluted by room air. Data suggest that for children, particularly children who have parents who smoke, PTS increases the occurrence of respiratory illness and chronic respiratory symptoms such as bronchitis, pneumonia, and coughing.[18,19] The health effects of PTS on respiratory symptoms and infection in adults have not been as well studied and the question remains controversial. Studies of the association between passive smoking and lung cancer in adults are also inconclusive; case-control and cohort studies do not uniformly indicate increased cancer risk. Despite conflicting and uncertain data, the International Agency for Research on Cancer of the World Health Organization, the National Research Council, and the U.S. Surgeon General have all concluded that involuntary smoking is a respiratory carcinogen.[18]

There is less evidence of adverse health effects for some of the other chemical compounds, listed above, partially because they occur infrequently in an indoor occupational environment. For example, though the effects of acute carbon monoxide poisoning by asphyxiation are known, health effects at low levels, and particularly from chronic exposure, are less well documented. Nitrogen dioxide occurs during combustion of gas during cooking and burning of pilot lights, and exposure is usually residential. The magnitude of respiratory illness resulting from exposure to NO_2 is usually small.[18]

Formaldehyde, VOCs, and Other Chemicals

There are numerous chemical compounds that may contribute to indoor air pollution. These include formaldehyde, carbon monoxide, and a category of complex mixtures of VOCs that are typically found in new buildings. Certain questions arise regarding these chemicals and their role in indoor air quality. We need to know, in a given instance, a) whether the chemicals are a proven cause of the illness; b) whether the levels at which the chemicals exist in the environment are known to cause the illness or symptoms; and c) what scientific methods have been used to measure the chemical levels, document symptoms, and prove causality.

Formaldehyde was used in urea-formaldehyde foam insulation (UFFI) until it was banned, but it also has numerous other sources in the home and office: particle board, paper products, floor coverings, and carpet backings are among the sources, albeit in very small amounts. Reported levels of formaldehyde in office buildings have ranged from 0.01 to 0.30 ppm. Levels that have been measured in buildings with no complaints were typically less than 0.1 ppm and reports of irritation of the eyes and upper respiratory tract have occurred at levels above 0.1 ppm. Formaldehyde has been associated with respiratory and neurobehavioral effects (at lower levels than OSHA permits), but this has not been proven. Published studies have been biased with regard to subject selection and data collection. Further investigation is needed.[14]

VOCs comprise many different compounds that have been identified in indoor air. As the technology for chemical analysis improves, we have the ability to identify trace amounts of these compounds. In a large-scale series of NIOSH investigations, 350 VOCs were identified in concentrations greater than 0.001 ppm (1 ppb). This does not mean that they exist in greater amounts than they did five years ago, or that exposure is greater, or that there is more danger from them, but simply that our techniques for finding them are better (see Chapter 8). In all cases, the measured levels of VOCs were within a factor of 100 of OSHA permissible exposure levels. Occupational standards are set at levels from 10 to 1,000 times below the expected no-effect levels in humans, so this means that measured levels of VOCs were always well below the generally believed no-effect levels for acute symptoms in humans.

With VOCs, as with other chemicals that may affect indoor air quality, we have conflicting or incomplete scientific evidence of toxicity at low levels. One study, for example, found that a small sample of healthy subjects, when exposed for a short time to VOCs, experienced subjective symptoms such as headache and general discomfort, but did not show any decreased performance on behavioral tests.[20] Two other studies did find that subjects exposed to organic solvents showed both cognitive deficits and psychological disturbances (similar to post traumatic stress disorder) on standardized tests. But nowhere in the studies is information given on the levels and intensity of exposure, information we need to compare the levels of the same chemicals in indoor air. These studies, then, should not be used as the basis for concluding that low-level VOCs cause neuropsychological disturbance.[21,22]

VOCs: A SUMMARY

The science of measurement has advanced rapidly, enabling us to detect indoor air contaminants at extremely low levels (ppb). When we conduct such measurements, we find hundreds of chemicals around us at all times at these levels. Residential air, building air, outdoor air, and air worldwide contains such materials. A 1989 report by the World Health Organization's (WHO) committee on Indoor Air Quality: Organic Pollutants, said, "...the indoor organic air pollutants as reported from several large surveys are similar in the distribution of concentrations in residential environments in several industrialized countries."[23] That same study discussed seventy-three chemicals commonly found in indoor air worldwide. Thus, it is quite easy to identify substances in indoor air. And their complex and frightening organic chemical names—hexane, formaldehyde, benzene, trichloroethylene, 1,1,1-trichloroethane, methylethylketone, and others—raise concerns for both those who believe themselves victimized by a building and those who are ultimately responsible for that building. By contrast, health effects of chemicals at these low levels are often nonexistent or unknown. For example, a common indoor air complaint is irritation of the nose and eyes. The WHO document described above notes, "In summary, organic compounds do produce mucosal irritation and other morbidity, though usually at orders of magnitude above the measured concentrations noted indoors."[23] Identified indoor air concentrations of substances are often thousands of times lower than those that are known to produce health effects. Workers in industries that use or manufacture those chemicals are frequently permitted exposures by OSHA of 100 to many thousands of times the levels found in buildings with no established untoward health effects. Critics of this comparison hark to differences in job requirements, such as the intense cognitive activities needed in offices and differences due to such things as "healthy worker effects" in which industries weed out sensitive workers. While these and other arguments may be valid, there is little proof that they are. And, to some extent, they become the rationalizations of those zealously committed to the belief that indoor air in modern buildings is uniformly bad and is producing significant health problems. Thus, frequently, levels of chemicals in buildings with worker complaints are no higher than customary background levels found in homes, shopping malls, and neighborhood restaurants. Invariably, they are vastly lower than permissible exposure levels in manufacturing facilities. This leaves investigators with data that identify chemicals but cannot correlate those levels of chemicals with the workers' symptoms. In most of those situations, the specific cause of complaints is never identified. Causes may be multifactorial, with perceptions and psychological factors playing a significant role.

INVESTIGATING INDOOR AIR COMPLAINTS

Correcting the causes of indoor air problems requires investigation, identification of specific causes, and implementation of a solution. The sheer nebulousness of these phenomena often makes a causal diagnosis difficult, and, consequently, remediation recommendations are fraught with uncertainties. Whether or not the planned course of action will take care of the problem is often unknown in advance and expensive remediation may proceed with no assurance that it will serve the intended purpose.

Dr. Morton Corn and others of the Johns Hopkins University School of Public Health compared two groups of people in two buildings in Washington, D.C. The first group complained of symptoms that they related to poor indoor air quality. The second, from a nearby building, had no such complaints. Workers in both buildings were given symptom questionnaires. Forty percent of those in the allegedly "sick building" returned the questionnaires; whereas 25% of those in the control "healthy building" returned theirs. The symptoms noted by both groups who returned questionnaires were identical: headaches and sinus problems predominating. Thus, the symptoms alleged are common complaints found in the population at large as well as among those who believe they are working in a "sick building." Whether the greater frequency of responses from those in the "sick building" (40% versus 25%) was due to actual physical problems caused by bad air in that building or to the perception that the building was producing a health hazard cannot be answered.[24] We do know, however, from an understanding of basic human psychology and from findings in the scientific literature, that a group of people who perceive a health threat are more likely to report symptoms than those who do not.[25-27] (See Chapter 5.)

This, then, highlights the central dilemma in "sick building" issues—the extraordinary difficulty of distinguishing between physical complaints due to perceived environmental threats versus illnesses genuinely caused by such factors. That distinction is frequently elusive, with complaints expanding beyond our ability to identify a physical cause. There is every reason to believe that many of the "sick building" complaints are based upon emotional responses to perceived hazards, group reinforcement of common fears, job dissatisfaction, and reporting biases as to actual chemical or other physical hazards.

DIVERSE PROFESSIONALS AND DISJOINTED RESPONSE

A significant impediment to the effective handling of such complaints arises because of the diversity of professionals involved in the relatively new area of indoor air quality. Environmental engineering firms may be prepared to measure substances in indoor air. Lacking toxicological or medical expertise, however, such firms may be ill-equipped to interpret the potential public health effects of their findings. Even less frequently are they able to deal with the complaints of specific individuals within that working environment. If they are not effective communicators or health professionals, they cannot respond effectively to the concerns of the workers. This may leave the employer or building manager with a set of data with no meaning, and no action plan. It is far easier to collect data than it is to interpret or act upon those data. Recent scientific developments make this particularly true.

Case Study

This case study from a recent consultation illustrates the profound problems in reconciling the perceptions with the realities. When children are the perceived threatened population, emotionality often holds sway over rationality.

In a school district in central Pennsylvania, an asbestos abatement program included removal of asbestos floor tiling with a petroleum-based chemical solvent. Following this, teachers and children noted "chemical" smells, sometimes quite intense, in certain classrooms. The school consulted an engineering firm that measured levels of specific VOCs, pronounced them "safe" because they were below occupational standards and departed, leaving teachers and, by now, a frantic parents group dissatisfied and more frightened than ever. After all, they had identified and measured chemicals and the basis for reassurance was a comparison with industrial workers, hardly an apt model for elementary school children, felt the teachers and parents.

As concerns mounted, so, too, did the range of symptoms. Headache, fatigue, and irritation were common. Other complaints included cough, increased frequency of colds, asthma, ear infections, upset stomachs, vomiting, diarrhea, and rashes.

NIOSH was called in. They performed a similar chemical survey and again declared the school safe—levels fell well below occupational standards.

By the time we got involved, the intense emotions and polarization had gripped this community. The school board was seen as not caring and covering up a potentially deadly situation. Teachers had mobilized in open revolt, as had the parent's group. Local newspapers and television stations had run stories emphasizing the hazards, the unknowns, the possibilities, and the childrens' fears. Attorneys were now beginning to enter the scene, offering to represent aggrieved parents in lawsuits on behalf of their children.

As we attempted to sort the scientific issues from the perceptions, it quickly became clear that doing so was essential, but not sufficient. Both had to be dealt with and treated if this situation were to be resolved. Several facts were apparent. First, there was a persistent smell of petroleum distillates in some classrooms. Second, there were no carcinogens of concern. Third, measured levels of chemicals were quite low, above the odor threshold, but vastly lower than thresholds of toxic levels either for exposed workers or for little children. Fourth, it was possible that levels of certain of the hydrocarbons were, at times, sufficiently high to produce some irritant symptoms. It was also quite clear that parents and teachers associated odors with toxicity and that they were absolutely convinced that the school was dangerous.

The best scientific explanation for the symptoms and complaints involved a combination of emotional response to odors, some irritant effects in sensitive individuals, and symptom intensification and recall based upon a perceived chemical threat. In other words, like the firefighters in Chapter 1, these teachers and students and their parents were primed emotionally to associate any and all symptoms and illnesses with that school environment.

Reassuring the parents and teachers that they were not being poisoned was an incomplete solution. Because the odors were so central to the symptoms and the perceptions, they had to be eliminated or, at least, reduced substantially. As it turned out, the asbestos remediation firm had saturated the concrete underlayment several inches deep with this petroleum-based product so odor remediation was no simple task. Ultimately, the resolution involved a combination of odor reduction and an intense educational effort. It was simply insufficient to compare these levels with permissible occupational limits. Parents had to be taught basic principles of toxicology, how we know that these levels of chemicals are not going to harm their children, why odor has little to do with toxicity, how and why symptoms have a variety of explanations, and even why fat accumulation of these chemicals (a fact which a doctor discussed with them) did not intensify the risk. Ultimately, solving such problems involves treating both the scientific aspects of toxicity as well as the perceptions.

CONCLUSION

Certain building-related diseases are readily definable—both the cause and the effect. Building-related symptoms and "tight" or "sick building" syndrome are another matter. Undoubtedly, instances of symptoms related to inadequate ventilation, specific chemical sources, and biological contamination do exist. What is unclear is how often and how many indoor air symptoms are actually affected by indoor air parameters. The literature is confusing and fraught with epidemiological confounders such as the use symptom questionnaires and the effects of both reporter and observer biases. These are clearly situations in which perceptions vastly overshadow the scientific data in intensity and level of certainty.

REFERENCES

1. Stolwijk, J.A.J., The "Sick Building" Syndrome, Department of Epidemiology and Public Health Yale University School of Medicine, New Haven, 1990 (unpublished report).

2. Woods, J.E., Drewry, G.M. and Morey, P.R., Office worker perceptions of indoor air quality effects on discomfort and performance, in *Indoor Air '87, Proceedings of the 4th International Conference on Indoor Air and Climate*, Seifert, B., Esdorn, H., Fischer, M., Ruden, H. and Wegner, J., Eds., Institute for Water, Soil and Air Hygiene, Berlin, 1987.

3. Woods, J.E., Cost avoidance and productivity in owning and operating buildings, *Occup. Med.*, 4, 575, 1989.

4. Fraser, D.W., Tsai, T.R., Ornstein, W., Parkin, W.E., Beecham, H.J., Sharrar, R.G., Harris, J., Mallison, G.F., Martin, S.M., McDale, J.E., Shepard, C.C., Brachman, P.S. and the field investigative team, Legionnaire's disease: description of an epidemic of pneumonia, *N. Engl. J. Med.*, 297, 1189, 1977.

5. McDade, J.E., Shepard, C.C., Fraser, D.W., Tsai, T.R., Redus, M.A., Dowdle, W.R., and the laboratory investigation team, Legionnaire's disease: isolation of a bacterium and demonstration of its role in other respiratory diseases, *N. Engl. J. Med.*, 297, 1197, 1977.

6. Hodgson, M.J., Frohlinger, J., Permar, E., Tidwell, C., Traven, N.D., Olenchock, S.A. and Karpf, M., Symptoms and microenvironmental measures in non-problem buildings, *J. Occup. Med.*, 33, 527, 1991.

7. Daniell, W., Camp, J. and Horstman, S., Trial of a negative ion generator device in remediating problems related to indoor air quality, *J. Occup. Med.*, 33, 681, 1991.

8. Brundage, J.F., Scott, R., Lednar, W.M., Smith, D.W. and Miller, R.N., Building-associated risk of febrile acute respiratory diseases in army trainees, *JAMA*, 259, 2108, 1988.

9. Colligan, M.J., The psychological effects of indoor air pollution, *Bull. N.Y. Acad. Med.*, 5, 1014, 1981.

10. National Institute for Occupational Safety and Health, *Guidance for Indoor Air Quality Investigations*, National Technical Information Service, Springfield, VA, 1987.

11. Finnegan, M.J., Pickering, C.A.C. and Burge, P.S., The sick building syndrome: prevalence studies, *Br. Med. J.*, 289, 1573, 1984.

12. Burge, S., Hedge, A., Wilson, S., Bass, J.H. and Robertson, A., Sick building syndrome: a study of 4373 office workers, *Ann. Occup. Med.*, 31, 493, 1987.

13. Skov, P., Valbjorn, O. and the Danish indoor climate study group, The sick building syndrome in the office environment: the Danish town hall study, *Environ. Int.*, 13, 339, 1987.

14. Letz, G., Sick building syndrome: acute illness among office workers: the role of building ventilation, airborne contaminants and work stress, *Allerg. Proc.*, 11, 109, 1990.

15. Kreiss, K., The epidemiology of building-related complaints and illness, *J. Occup. Med.*, 4, 575, 1989.

16. Donnell, H.D., Bagby, J.R., Harmon, R.G., Crellin, J.R., Chaski, H.C., Bright, M.F., Van Tuinen, M. and Metzger, R.W., Report of an illness outbreak at the Harry S. Truman state office building, *Am. J. Epidemiol.*, 129, 550, 1989.

17. Alexander, R.W. and Fedoruk, M.J., Epidemic psychogenic illness in a telephone operator's building, *J. Occup. Med.*, 28, 42, 1986.

18. Samet, J.M., Marbury, M.C. and Spengler, J.D., Health effects and sources of indoor air pollution: part 1, *Am. Rev. Resp. Dis.*, 136, 685, 1987.

19. Samet, J.M., Marbury, M.C. and Spangler, J.D., Respiratory effects of indoor air pollution, *J. Allergy Immunol.*, 79, 685, 1987.

20. Otto, D., Molhave, G., Rose, G., Hundell, H.K. and House, D., Neurobehavioral and sensory irritant effects of controlled exposure to a complex mixture of volatile organic compounds, *Neurotoxicol. Teratol.*, 12, 649, 1990.

21. Morrow, L.A., Ryan, C.M., Goldstein, G. and Hodgson, M.J., A distinct pattern of personality disturbances following exposure to mixtures of organic solvents, *J. Occup. Med.*, 31, 743, 1989.

22. Morrow, L.A., Ryan, C.M., Hodgson, M.J. and Robin, N., Alterations in cognitive functioning after organic solvent exposure, *J. Occup. Med.*, 32, 444, 1990.

23. Indoor Air Quality: Organic Pollutants: Report on a WHO Meeting, Berlin (West) 23-27 August 1987, WHO Regional Office for Europe, Copenhagen, 1989.

24. Corn, M., personal communication, 1991.

25. Bond, S., Beringer, G.B., Kundin, W.D., et al., Epidemiologic Problems Related to Medical Coverage of New Diseases, presented at annual meeting of the American Public Health Association, Dallas, November 1983.

26. Hopwood, D.G and Guidotti, T.L., Recall bias in exposed subjects following a toxic exposure incident, *Arch. Environ. Health,* 43, 234, 1988.

27. New Jersey Department of Health, Environment Health Hazard Evaluation Program (prepared in cooperation with Atlantic County Health Department), *Health Survey of the Population Living Near the Price Landfill, Egg Harbor Township, Atlantic County,* New Jersey Department of Health, New Jersey, 1983, 29.

14
ASBESTOS

For nearly 20 years, asbestos has occupied center stage in the environmental risk and litigation arenas. Most people are well aware of the legacy of asbestos-covered workers, whose chronic lung diseases and cancers present an unhappy chapter in American industrial history. The ubiquitous use of asbestos makes it a substance with which most of us have become familiar. Workers in white suits and respirators laboring in plastic bubbles to rid commercial buildings, schools, and homes of the hazardous material are familiar sites, symbolic of the prototypical environmental toxic threat. News of its pervasive existence in public buildings, particularly schools, has created widespread fear, particularly over the safety of vulnerable children.

While certain aspects of the asbestos story are soundly rooted in unimpeachable science, other elements are less so. As communities are forced to confront the sheer enormity and fiscal monstrosity of the asbestos removal tasks, risks are increasingly being weighed against the benefits of removal. Such risk assessment has required intense reappraisal of the scientific principles underlying abatement decisions. Specifically, two major areas of uncertainty—the effects of low-level exposure and the differential risks of different fiber types—have surfaced as the critical issues in this new risk assessment. As these issues emerge and become central to the debate, personal views, perceptions, and vested interests among scientists and regulators have propelled the asbestos debate to a new level of intensity and blatant acrimony. Nowhere is the line between scientific realities, toxic perceptions, and scientific activism more sharply drawn than in this discussion of the hazards of asbestos and how to manage them.

ASBESTOS AND LUNG DISEASE: THE SCIENCE

Asbestos is the name given to a group of naturally occurring mineral silicates that are separable into fibers. The various forms of asbestos are differentiated from one another by the size and shape of the fiber. Serpentine asbestos (predominantly crysotile) is curly. Amphiboles (crocidolite and amosite are the most important) are long and thin.

Certain forms of asbestos are now known to produce several diseases in human beings. The first is asbestosis, a disease characterized by scarring of the lung with resulting loss of elasticity. This restrictive lung disease ultimately limits the ability to move air in and out of the lungs and to transport essential gases, oxygen and carbon dioxide, across the alveolar membrane. In its severe form asbestosis can be fatal. Asbestosis has been recognized since the early part of the 20th century.[1-5] A disease related to dose, it is strictly occupational in origin and primarily seen in heavily exposed individuals: miners, insulators, and pipefitters. The definition of asbestosis at the lower end of the exposure spectrum has been debated and ranges from full-blown impairing disease to minor changes on X-rays. It is unlikely that anyone would have classified these minor changes as asbestosis had it not been for the pressures of compensation and litigation. Claims for damages have pressed inexorably for a definition of asbestos-related disease at this subclinical extreme.

Because exposures are lower today, clinically important asbestosis is not expected to be a major public health issue in the future. Far more relevant are the malignancies that have been associated with asbestos exposure. These have extended the debate beyond heavy occupational exposures to less exposed workers (brake repairmen, custodians, and others) and have raised the specter of general public exposure and a widespread public health risk. These malignancies have produced the Asbestos Hazard Emergency Response Act of 1986 (AHERA) and has led the call for eradication of every last fiber of asbestos.

Malignant mesotheliomas are fatal tumors that develop in mesothelial cells in the pleura, pericardium, and peritoneum. They are quite rare, only 1,648 in the United States recorded between 1973 and 1984.[6] The connection between mesotheliomas and asbestos was first described in South African miners whose predominant exposure was to crocidolite asbestos.[7] Since then, mesotheliomas have been associated predominately with crocidolite-exposed workers, and many now believe that they either do not develop at all from crysotile asbestos or, if they do, the risk is diminished by comparison to the risk from the amphiboles.[8-10]

By the mid 1950s the relationship between asbestos exposure and carcinomas of the lung was reported.[11] The major recognition of this relationship took place in the 1960s when Selikoff and others reported the results of an extensive study of an asbestos-exposed cohort of insulators. A particularly concentrated industrial use of asbestos had occurred in the 1940s when hundreds of ships in frenzied preparation for war were outfitted with asbestos-insulated pipes. Those shipbuilders and pipefitters, heavily exposed to asbestos, ultimately became the cohort for the studies that linked, beyond question, heavy asbestos exposure to lung cancers.[12,13]

There is no dispute over the relationship between heavy exposures to specific asbestos types and crippling or fatal lung diseases. Asbestosis, mesothelioma, and lung cancer are all caused by asbestos. There the science is quite clear. What is less clear, and now the subject of substantial debate, is the effect of low-level asbestos exposure and whether certain fiber types, specifically crysotile, are less hazardous, perhaps even harmless.

Both of these questions have enormous implications in public policy, economics, and compensation for claimed asbestos-related disease. This is so because major exposures have now been controlled and today's exposures are far lower. The heaviest are in miners, but new drilling machinery and protective equipment have lessened that. Other occupational exposures such as that associated with brake repairs are less. Still others such as custodial and electrical work in asbestos-containing buildings are even less. Even lower are the low-level environmental exposures associated with ambient air in asbestos-containing buildings. Thus, current exposures bear little relationship to the heavy exposures of the past. In addition, the vast majority of asbestos remaining in our buildings is chrysotile. If chrysotile is less hazardous, abatement becomes less necessary.

Low Levels of Asbestos

The cohorts of workers who suffered substantial risks of asbestos-related cancers worked in environments white with the dust containing hundreds of fibers per cubic centimeter. In 1971 the OSHA reduced the acceptable exposure level to fibers greater than 5 μm in length to five fibers per cc. In 1986 this standard was further reduced to 0.2 fibers per cc. Levels found in indoor air, even in rooms with damaged asbestos insulation are generally less than 1/100th of that amount. Whether very small amounts pose some risk that, when spread over a population of greater than 200 million, may result in 100 or 1,000 cancers cannot be answered fully. The arguments here—threshold versus no threshold, linear versus other forms of quantitative risk assessment—are the same for asbestos as they are for other toxins. These have been discussed in Chapters 9 and 10. There are, however, certain scientific clues that argue against the proposition of some risk at any level. First of all, asbestos is not generally considered to be genotoxic. It is, in fact, only one of two Class 1 (known to cause cancer in humans) carcinogens which is not. The arguments in favor of a no-threshold theory generally rely upon genotoxicity and the ability to produce mutation at any level to support that theory. Asbestos does not exhibit such behavior.[14] Second, because they are so rare and so strongly associated with asbestos, mesotheliomas are both a sensitive and specific outcome marker. If low-level asbestos exposure were carcinogenic, one would have expected to have seen an excess in mesotheliomas in the population at large from indoor air and other background exposure. Despite the fact that we customarily breathe in 1,000,000 or so asbestos fibers each year, numerous epidemiological studies have found no notable risk in the general population.[6,15,16] Thus, although one can never disprove some risk even of a single fiber, the belief in such a risk is an article of faith unsupported by scientific data.

Fiber Type

The second major scientific issue and public policy battleground is over the role of fiber type in the genesis of mesotheliomas and lung cancers. The central question is whether chrysotile asbestos is either less carcinogenic than the amphiboles or, perhaps, not carcinogenic at all. Since 90-95% of our commercial asbestos is chrysotile, this question has profound financial and public policy significance.

In the early days of the asbestos-cancer story, there was little attention, interest, or even ability to evaluate separately different fiber types. As electron microscopic techniques improved and researchers began to probe the specific mechanisms of asbestos—mediated carcinogenesis, fiber-response specificities emerged. It became clear, for example, that the greater the ratio of fiber length to its diameter the more hazardous the fiber. It also became apparent that shorter fibers (under 5 μm in length) were less hazardous than longer ones.[9] In these regards, the science is clear; there is little argument. Argument abounds, however, over the unanswered and practically unanswerable questions: Do asbestos fibers become harmless at some length? If so, exactly what is that length?

With this recognition of variations in hazardous propensity as a function of fiber size came an obvious question: Are all chemical forms of asbestos equally hazardous? Chrysotile asbestos is quite different physically, structurally, and chemically from the amphibole forms. One might, therefore, expect that it could behave differently biologically. In animal models, both chrysotile and the amphiboles produce mesotheliomas when they are injected directly into the peritoneum.[17] Some have speculated, therefore, that differential alveolar penetration rather than inherent carcinogenicity may account for the apparent reduced risk of chrysotile noted in epidemiological studies of exposed human beings.[8] Chrysotile is curly and clumps together. It has, if you will, a "softer" appearance than the fibrous amphiboles, which break off into sharp, sharp, needle-like fragments. The latter appear to penetrate the lung tissue more effectively than does the chrysotile. Studies have also suggested that chrysotile is cleared more readily than the amphiboles, thereby less often reaching the deep recesses of the respiratory tree. Also, the unique chemical composition of chrysotile seems to make it more soluble in tissue, less likely to remain permanently.

Supporting this soft evidence of chrysotile's reduced hazard are experimental and epidemiological data. The most compelling evidence that chrysotile may be less carcinogenic than are the amphiboles comes from a large number of epidemiological studies and supported by a series of autopsy studies. In Quebec largely chrysotile mining towns with generations of exposure have few mesotheliomas attributable to nonoccupational exposures.[10] Miners and millers in Quebec with largely chrysotile exposure have developed few mesotheliomas.[18,19] Those that have had some concomitant amphibole exposure, giving rise to strong argument that mesotheliomas in these cohorts requires amphibole exposure.[8,10] A British cohort study examining workers exposed to chrysotile alone at levels of 0.5 to 1.0 fibers per cc has shown no excess deaths from lung cancer or mesotheliomas.[20] Many other epidemiological studies and reports have confirmed these findings.[15,18,21,22]

Supporting these epidemiological observations are studies of pulmonary tissue from individuals who died from malignant mesotheliomas. Studies have examined the prevalence of chrysotile and amphibole fibers in such individuals. They have consistently shown that the concentrations of amphiboles, but not of chrysotile, exceeds that of controls.[10,23] Putting together these pathological findings and the various epidemiological studies, many researchers have concluded that chrysotile is less hazardous than the amphiboles. Recently, Dr. Richard Doll, summarizing the results of a international conference, concluded:

> That pure chrysotile does not cause mesothelioma is strongly suggested by the low incidence of the disease in groups of men and women occupationally exposed to only chrysotile (with or without some contamination by tremolite) and by the results of tissue analyses which have repeatedly shown that the lungs of people who have died of mesothelioma contain very little (if any) more chrysotile than the lungs of those who have died of non-asbestos-related diseases....[9]

Because of the profound financial and public policy implications of asbestos in commercial buildings, including schools, the EPA and Congress commissioned an analysis by a prominent panel. The Health Effects Institute Asbestos Research (HEI-AR) published its findings in 1990 in a report entitled "Asbestos in Public and Commercial Buildings: A Literature Review and Synthesis of Current Knowledge."[24] The purpose of this report was to analyze the state of the science with regard to a number of issues: identification of concentrations of asbestos in public and commercial buildings; concentrations of fibers to which general occupants, maintenance workers, custodial workers, abatement workers and others are exposed and the potential health effects from those exposures; the impacts of various remediation methodologies upon exposure and health risk; the significance of each form of asbestos regarding health effects and its implications for different remediation options. The report was a comprehensive review of the state of the science regarding relatively low-level exposures. The panelists were careful to highlight the scientific uncertainties created by insufficient data or fundamentally unanswerable questions. Some of their key observations included conclusions about fiber length, fiber diameter, fiber type and risk assessment. They concurred that fiber length above 5 μm is appropriate for risk assessment considerations, but pointed out that whether shorter fibers are noncarcinogenic in human beings is not known. They noted that thin fibers seem more hazardous than thicker fibers. They acknowledge the apparent lower risk of crysotile asbestos, but conclude that the issue is sufficiently unsettled scientifically to warrant a separate risk assessment. Finally, their risk assessment is shown in Table 14.1. Critical assumptions and uncertainties underlying this risk assessment were delineated in the report. First, a linear extrapolation (see Chapter 10) was performed from effects of heavy industrial exposures (10 f/cc).

Second, the assessment was based upon fibers greater than 5 μm in length identified by transmission electron microscopy (TEM). Third, assumptions were made to extrapolate from the tested buildings to buildings generally. This extrapolation may have been overly conservative or insufficiently conservative. Finally, there exist uncertainties about aspects of measuring techniques including sample preparation, analytical sensitivity, and measurement errors.

Recognizing the limitations of the data and the necessary assumptions and extrapolations, this risk assessment suggests rather low risk from buildings containing asbestos. The cancer risk from school attendance, for example, is approximately the same as the asbestos-related risk from outdoor air exposure. A significant conclusion of this report was that "although public concern over asbestos in buildings has focused primarily on potential risks to general building (C1) occupants, there does not appear to be sufficient grounds for arbitrarily removing intact ACM (asbestos-containing material) from well-maintained buildings...."

Table 14.1
Estimated Lifetime Cancer Risks for Different Scenarios of Exposure to Airborne Asbestos Fibers

Conditions	Premature Cancer Deaths (Lifetime Risks) per Million Exposed Persons
Lifetime continuous outdoor exposure	
• 0.00001 f/ml from birth (rural)	4
• 0.0001 f/ml from birth (high urban)	40
Exposure in a school containing ACM, from age 5-18 years (180 days/year, 5 hours/day)	
• 0.0005 f/ml (average)	6
• 0.005 f/ml (high)	60
Exposure in a public building containing ACM age 25-45 years (240 days/year, 8 hours/day)	
• 0.0002 f/ml (average)	4
• 0.002 f/ml (high)	40
Occupational exposure from age 25 to 45	
• 0.1 f/ml (current occupational levels)	2,000
• 10 f/ml (historical industrial exposures)	200,000

Source: Health Effects Institute-Asbestos Research, *Asbestos in Public and Commercial Buildings: A Literature Review and Synthesis of Current Knowledge*, Health Effects Institute-Asbestos Research, Cambridge, MA, 1991, 1-11.

This report was a thoughtful, balanced summary of the state of scientific knowledge. Limitations of the data, scientific uncertainties, and necessary assumptions were carefully delineated. Despite explicit discussion of these limitations, two panel members rejected the report and refused to sign off on it. Each felt that the report was biased, either underestimating (William J. Nicholson, Ph.D.) or overestimating the risks (J. C. Wagner). Quite clearly, these represented certain intense opinions by those panelists who could not accept a report that failed to support their opinions.

THE PUBLIC POLICY REGULATORY SCIENCE DEBATE

This HEI-AR report and the well-reasoned, calm letters of the dissenting voices are genteel antagonists compared with the firestorm of current acrimony. Resistant to new debates about fiber types, fiber lengths, and dose are the institutions, deeply entrenched in asbestos dogma and past beliefs that any amount of any type is unacceptably hazardous. These areas of resistance are found in the legal arena where asbestos claims are decided, among companies that depend upon asbestos removal for their existence, and in segments of the public health and regulatory arenas in which physicians and scientists, steeped in asbestos dogma, are angry and vocal over a perceived "revisionist" scientific philosophy.

Dr. Selikoff and his associates (including Dr. Nicholson, who protested the above report) are the most outspoken. They have strongly asserted that individuals such as custodians, construction workers, firefighters and others having relatively low-level exposures will become the "third wave" of asbestos claimants, potentially numbering in the many thousands. Their position in this regard is not based in science, but rather upon their opinions. They have, however, been so outspoken in this viewpoint that they seem unable and unwilling to discuss the science pertaining to any of these "revisionist" issues. They hold their own conferences in which "revisionists" are not invited to participate. Quoted in a report in *Science,* Dr. Bruce Case, director of the Center for Environmental Epidemiology at the University of Pittsburgh, described his feelings about the June 1990 meeting entitled "The Third Wave of Asbestos Diseases: Exposure to Asbestos in Place, Public Health Control." He said,"... it was like nothing I've ever attended before. This is the first meeting I'd ever gone to where I got the feeling that the whole thing was a kind of political setup."[25]

The beginning discussions of the "revisionist" concepts began in the late 1970s when Dr. Hans Weill and Christopher Wagner (another dissenter in the HEI-AR report) first argued that fiber type was relevant to risk.[24-28]

Discussions intensified and sparks between "revisionists" and "third wavers" flew in the mid to late 1980s. In 1988 the "revisionists" held a conference at Harvard in which the conclusion was reached that the risk of asbestos in place in buildings was quite small. The passions and the debate have intensified in the past few years, now that enormous fiscal issues are immediately at hand.

A *New York Times Magazine* article recently featured this struggle between the "revisionists" and the "third wavers." In his article, "The Asbestos Mess," Joseph Hooper effectively presents the current sides of the debate.[29] What is not clearly articulated is the reason for this clash of opinions. At stake is public confidence in science and the scientific process itself. The scientific process uncovers facts. The better the investigation, the stronger those facts and the more widespread their acceptance in the scientific community. It isn't that scientists cannot agree upon facts. Rather, public policy inevitably brings in an admixture of fact and belief. Problems in communication and public confusion arise when scientists forget how to distinguish fact from belief. When they represent their beliefs as facts they appear confused for other scientists will then have a different set of "facts." This wide disparity in facts cannot exist, but there can be many opinions. Some scientists, intensely involved in policy making, forget that distinction. Fact is what has been proven to be true. Policy relies upon an admixture of the known, the possible, and the prudent. What we know about asbestos (scientific fact) is that certain industrial workers—insulators and shipbuilders—heavily exposed to asbestos in the 1940s and 1950s, developed serious illnesses including lung cancers and mesotheliomas from those exposures. About this the science is clear and Dr. Selikoff's statement, "There is no issue, no debate about asbestos," is irrefutable. What is less certain scientifically, creating substantial issue and debate, is whether all forms of asbestos are equally dangerous and whether small quantities are harmful just because large quantities are. Here reputable scientists disagree because facts are either unclear or nonexistent. They may have strong opinions in one direction or another, but they do not have irrefutable scientific data. In the face of this uncertainty, public policy takes over and statutes are enacted such as AHERA, the act that required evaluation and development of a management plan for handling asbestos in schools. The problem is not that scientists can't agree. The problem is that they have been unable or unwilling to identify clearly, so that the public might understand, this line between established scientific knowledge and public policy.

LITIGATION

Most readers are aware of the recent history of asbestos and its diseases. Tens of thousands of workers filed both workers' compensation claims and tort suits against the manufacturers, claiming that they were made ill by asbestos. Asbestosis, lung cancer, and mesothelioma were the predominant diseases claimed, but others including esophageal and colon cancer have been claimed as well. Companies and their insurers spent hundreds of millions of dollars defending and settling these claims. Some, such as John Mansville and the Keene Corporation, reorganized through bankruptcy because of the litigation. Still others folded and went out of business.

The claims involving the exposures of the 1940s were similar to the environmental excesses generally. The science was clear, relationships distinct. A 1940s shipbuilder or insulator who couldn't breathe and had certain physical and laboratory findings could be confidently diagnosed as having asbestosis. Even then, arguments and denials would arise over whether the disease was caused by smoking or by the asbestos. Sometimes that distinction was clear, sometimes not. Similarly, lung cancers in this population were readily found causally related to the exposure since excess risks were enormous.

Today, things have changed as we have moved into the era of more subtle causal attributions. The first wave of asbestos claimants brought with them enormous national attention and worldwide visibility. It left in its wake a billion-dollar industry of attorneys (both plaintiff or petitioner and defense), claims handlers, insurance specialists, scientific and medical experts and scientists, and even entire university departments devoted to this disease. A train of this size does not stop rolling once most of its passengers have disembarked. It continues on its way looking for new problems and invariably finding them whether or not they exist.

Today, claims are filed based upon less clear diseases and causal relationships. A brake repairman with far less exposure than the insulators ever had may claim that a mild lung disorder evidenced only by subtle X-ray alterations has a compensable asbestos-related disease. There the disease is in question; the amount of exposure is in question; even whether such an exposure could ever produce asbestosis is uncertain. At the extreme of even lower exposures are such current claimants as janitors and other custodial workers who claim that indoor air contaminated by asbestos fibers produced diseases.

When we move beyond the suffocating disease of major asbestosis, we find in these new claims far less evidence and more subtle clinical problems. Whether these even represent a disease at all is frequently in question, as we see in the following example.

A recent example of an outbreak of workers' compensation claims of asbestosis later studied more scientifically and reclassified occurred in a group of tireworkers. During 1986 over 700 tireworkers at a specific site were evaluated for asbestosis with a variety of screening methods. Following the evaluation, 440 filed claims for asbestosis based upon X-ray changes, generally quite subtle. Concerned about the enormous prevalence of diagnosed asbestos-related disease in these workers (440 out of 700 to 750 or approximately 60%), Reger and others carried out a controlled study to reevaluate the findings.[30] They took the X-rays and submitted them in a blinded fashion to three radiologists. A consensus of their findings found that only 11 workers or 2.5% of the total population had a reasonable likelihood of having asbestos-related disease. This represented an astounding error rate in the initial diagnoses (40-fold) that formed the bases for these claims. Extrapolated generally to subtle X-ray changes claimed to be asbestos-related, this could represent tens of thousands of erroneous diagnoses and erroneous claims for compensation.

In another example, Oliver and colleagues studied school custodians who had ill-defined asbestos exposure, but clearly at low levels. In a study designed to confirm a belief that these workers were truly at risk, she and her colleagues found "changes" on X-rays consistent with asbestosis—very much as occurred in the tireworkers.[31] In her work the existence of these findings has not been confirmed through standard double-blind multiple reviewer methodologies. Moreover, even the definition of what constitutes a "disease" becomes an issue. If, for example, a pleural plaque (scar of the pleural lining) exists, Dr. Oliver and her colleagues classify that as an asbestos-related disease. Yet there is little evidence that such plaques cause harm or that they are any more significant than a healed cut of the finger with its resultant scar.

REFERENCES

1. Pancost, H.K., Miller, T.C. and Landis, H.R.M., A roentgenologic study of the effects of dust inhalation upon the lungs, *Trans. Assoc. Am. Physicians*, 32, 97, 1917.

2. Hoffman, F.L., Mortality from respiratory diseases in dusty trades, (inorganic dust), *Bull. U.S. Bur. Labor Statistics*, 231, 176, 1918.

3. Cooke, W.E., Pulmonary asbestosis, *Br. Med. J.*, 2, 1029, 1927.

4. Cooke, W.E., Fibrosis of the lungs due to the inhalation of asbestos, *Br. Med. J.*, 2, 579, 1929.

5. Merewether, E.R.A., The occurrence of pulmonary fibrosis and other pulmonary affections in asbestos workers, *J. Ind. Hygiene*, 12, 198-222; 239-257, 1930.

6. Connelly, R.R., Spirtas, R., Meyers, M.H., Percy, C.L. and Fraumeni, J.F., Demographic patterns for mesothelioma in the United States, *J. Natl. Cancer Inst.*, 78, 1053, 1987.

7. Wagner, J.C., Sleggs, C.A. and Marchand, P., Diffuse pleural mesothelioma and asbestos exposure in the North-West Cape Province, *Br. J. Ind. Med.*, 17, 260, 1960.

8. Mossman, B.T., Bignon, J., Corn, M., Seaton, A. and Gee, J.B.L., Asbestos: scientific developments and implications for public policy, *Science*, 247, 294, 1990.

9. Doll, R., Mineral fibers in the non-occupational environment: concluding remarks, *IARC Sci. Publ.*, 90, 511, 1989.

10. McDonald, J.C. and McDonald, A.D., Asbestos and carcinogenicity, *Science*, 249, 844, 1990.

11. Doll, R., Mortality from lung cancer in asbestos workers, *Br. J. Ind. Med.*, 12, 31, 1955.

12. Selikoff, I.J., Churg, J. and Hammond, E.C., Asbestos exposure and neoplasia, *JAMA*, 188, 22, 1964.

13. Selikoff, I.J., Hammond, E.C. and Churg, J., Asbestos exposure, smoking and neoplasia, *JAMA*, 204, 104, 1968.

14. Shelby, M.D., The genetic toxicity of human carcinogens and its implications, *Mutat. Res.*, 204, 3, 1988.

15. McDonald, A.D. and McDonald, J.C., in *Asbestos Related Malignancy*, Antman, K.H. and Aisner, J., Eds., Grune and Stratton, Orlando, 1987, 31-55.

16. Enterline, P.E. and Henderson, V.L., Geographic patterns for pleural mesothelioma deaths in the United States, 1968-81, *J. Natl. Cancer Inst.*, 79, 31, 1987.

17. Jaurand, M.C., Fleury, J., Monchaux, G., Nebut, M. and Bignon, J., Pleural carcinogenic potency of mineral fibers (asbestos attapulgite) and their cytotoxicity on cultured cells, *J. Natl. Cancer Inst.*, 79, 797, 1987.

18. Churg, A., Chrysotile, tremolite, and malignant mesothelioma in man, *Chest*, 93, 621, 1988.

19. McDonald, J.C., Liddell, F.D., Gibbs, G.W., Eyssen, G.E. and McDonald, A.D., Dust exposure and mortality in chrysotile mining, 1910-75, *Br. J. Ind. Med.*, 37, 11, 1980.

20. Newhouse, M.L. and Sullivan, K.R., A mortality study of workers manufacturing friction materials: 1941-86, *Br. J. Ind. Med.*, 46, 176, 1989.

21. Gardner, M.J., Winter, P.D., Pannett, B. and Powell, C.A., Follow up study of workers manufacturing chrysotile asbestos cement products, *Br. J. Ind. Med.*, 43, 726, 1986.

22. Sebastien, P., McDonald, J.C., McDonald, A.D., Case B. and Harley, R., Respiratory cancer in chrysotile textile and mining industries: exposure inferences from lung analysis, *Br. J. Ind. Med.*, 46, 180, 1989.

23. McDonald, J.C., Armstrong, B., Case, B., Doell, D., McCaughey, W.T., McDonald, A.D. and Sebastien, P., Mesothelioma and asbestos fiber type. Evidence from lung tissue analyses, *Cancer*, 63, 1544, 1989.

24. Health Effects Institute-Asbestos Research, *Asbestos in Public and Commercial Buildings: A Literature Review and Synthesis of Current Knowledge*, Health Effects Institute-Asbestos Research, Cambridge, MA, 1991.

25. Stone, R., No meeting of the minds on asbestos, *Science*, 254, 928, 1991.

26. Weill, H., Rossiter, C.E., Waggenspack, C., Jones, R.N. and Ziskind, M.M., Difference in lung effects resulting from crysotile mine and mill ex-workers, *Arch. Environ. Health*, 38, 54, 1979.

27. Weill, H., Hughes, J. and Waggenspack, C., Influence of dose and fiber type on respiratory malignancy risk in asbestos cement manufacturing, *Am. Rev. Resp. Dis.*, 120, 345, 1979.

28. Wagner, J.C., Berry, G. and Pooley, F.D., Mesotheliomas and asbestos type in asbestos textile workers: a study of lung contents, *Br. Med. J.*, 285, 603, 1982.

29. Hooper, J., The asbestos mess, *New York Times Magazine,* November 25, 1990.

30. Reger, R.B., Cole, W.S., Sargent, E.N. and Wheeler, P.S., Cases of alleged asbestos-related disease: a radiologic re-evaluation, *J. Occup. Med.*, 32, 1088, 1990.

31. Oliver, L.C., Sprince, N.L. and Greene, R.E., Asbestos related radiographic abnormalities in public school custodians, *Toxicol. Ind. Health*, 6, 629, 1990.

15 LEAD

The headlines and media "statistics" are compelling. They tell us that lead is the number one threat to our children, that middle-class children with even low lead levels are six times more likely to have reading disabilities and seven times more likely to drop out of high school.[1] Lead, it is said, is slowly killing or causing brain damage to millions of our children. No one, it seems, is safe from "the silent hazard."

Thus, the story of lead toxicity has in recent years taken on the qualities of soap opera drama. It represents the struggle between classes: residents of the inner city versus wealthy landlords. In its recent iteration, its victims are children, innocent, sympathetic victims. Sociologists have claimed that poverty and educational disadvantage can be blamed solely upon lead poisoning. Politicians are running on lead platforms. New lead abatement industries have sprung up seemingly overnight, selling overpriced remediation services, capitalizing upon the fears of newly urbanized middle-and-upper class parents. When it comes to perceptions versus science, to the cold analysis of data versus impassioned concern for these "victims," the "winner" is obvious. Solid facts cannot begin to compete with the images of poisoned children, their opportunities removed by corporate greed and callous disregard of others. Fueling the perceptions and driving regulation with enormous ferocity are a set of "soft facts" about lead and lead poisoning. These soft facts have moved the Centers for Disease Control (CDC) to reduce the standard for lead poisoning in children from 25 μg/dl to 10 μg/dl and have stimulated the declaration that "approximately one of every nine American children may be lead poisoned."[1]

That lead at high levels is a powerful neurotoxin is well known and not subject to general controversy. Where solid science leaves us is in the realm of lower level exposures and more subtle health effects. The question in dispute and not answered scientifically is whether or not children are being adversely affected by lead at body levels well below the accepted threshold of lead poisoning. Unlike the situation with an agent such as Alar (Chapter 16), in which perceptions have been based solely upon animal models, the lead literature does not suffer from a dearth of human data. Rather, scientific uncertainties arise because of the limitations in experimental design and interpretation. The recent human lead literature exploring low-level exposures suffers from all of the methodological problems in experimentation discussed in Chapters 3 and 5. Vast resources are being directed to these questions. Are lower levels of lead associated with neurological impairment? Does the excessive lead burden during the period of central nervous system maturation affect the intellectual and behavioral development of children when the burden is not associated with overt signs of neurotoxicity? Is lead related to deficits in the cognitive abilities or educational attainment of children at low levels of exposure? These issues have been explored by many investigators, but because of profound experimental impediments, are not fully answered and may never be. Today, however, public policy determinations are rapidly taking place, based upon the most dire of the controversial findings.

WHERE LEAD COMES FROM

The sources of lead in our environment have been explored quite thoroughly. Here the science is clear, for lead can be measured in various environmental media. Lead is and will continue to be an important metal with many industrial uses, the principal one being the electric battery industry. However, leaded gasoline combustion in vehicles has accounted for as much as 90% of the total anthropogenic sources of environmental lead. From 1975 to 1984, lead in gasoline decreased 73%, while lead in the ambient air decreased 71%. It should not be thought, however, that all human exposure to lead is from the inhalation of airborne lead. For about four decades now, more than 100,000 to 200,000 tons of lead per year have been emitted from automobile exhausts in the United States. This lead is deposited in and fixed to soil, taken up by plants, and then enters our water sources. Currently, the major sources of human exposure to lead include food and beverage (because of lead uptake by food sources and its presence in drinking water), lead paint, and respirable and edible lead dust.[2,3]

In 1986 the U.S., with the enactment of Congress Section 118(f), Superfund Amendments and Reauthorization Act (SARA), directed the ATSDR to quantify the contributions of various sources of lead to childhood exposure. According to the ATSDR, significant sources of lead in childhood exposure include lead in paint, dust, soil, and drinking water. Approximately 6 million U.S. children less than 7 years old reside in the oldest housing, with highest exposure risk due to leaded paint. About 2 million in deteriorated units are at particularly high risk for exposure, with 1.2 million children in the oldest, most deteriorated housing estimated to have blood lead (PbB) levels above 15 μg/dl. Soil and dust lead are potential sources of exposure for 6 million to 12 million children. Residential tap water lead is a measurable source for 3.8 million children, of whom the EPA estimates 240,000 have water-specific exposures at toxic levels.[2] Leaded gasoline combustion mainly in past years has produced, and will continue to produce into the 1990s, significant numbers of exposed children with toxicologically elevated PbBs. For 1990, 1.25 million children have had their PbBs fall below 15 μg/dl as a result of the decrease in leaded gasoline. Food lead can cause significant exposure in certain cases. Table 15.1 below provides estimates of the contribution of these sources.[4]

Table 15.1 Estimated Numbers of Children Exposed to Sources of Lead

Source	Number of Children* in Millions
Leaded Paint †	12.0
Leaded gasoline	5.6
Stationary sources (factories, smelters)	0.233
Dust\Soil	5.9 - 11.0
Water/Plumbing §	10.4
Food	1.0

* Numbers in the table are not additive because children are usually exposed to multiple sources of lead in the environment.

† Includes children in oldest housing: 5.9 million; children in deteriorated housing: 1.8 million to 2.0 million.

§ Residential service lead connection, 4.8 million; leachable lead solder, 1.8 million; water lead level >20μg/ml, 3.8 million.

Perhaps the most significant source of lead for childhood exposure is dust. Although dusts are of complex origin, they may be placed conveniently into a few categories relating to human exposure: household dusts, soil dust, street dusts, and occupational dust. In each case, the lead in dust arises from a complex mixture of fine particles of soil, flaked paint, and airborne particles of industrial or automotive origin. Dust is deposited on window sills from outdoor sources and from the friction created by windows moving up and down the lead-painted tracks. Dust particles characteristically accumulate on exposed surfaces and are trapped in the fibers of clothing and carpets.[5] Children, especially preschool children, need only to touch lead dust or a toy or bottle that has had lead dust settle upon it and then put their fingers or thumbs in their mouth or pick up a piece of food to eat to ingest lead dust. It is this ingestion of dust particles during eating and playing, rather than inhalation, that appears to be the greatest problem for children. Similarly, children playing outside are exposed to lead dust deposited on the sidewalk, in soil and sand, and on their outdoor toys. To be exposed to lead, they need only to play with their hands on the sidewalk and then put any part of their hands in their mouth.

According to the EPA, the lead content of dusts figures prominently in the total lead exposure picture for young children.[5] Lead in aerosol particles deposited on rigid surfaces in urban areas (such as sidewalks, porches, steps, and the like) does not undergo dilution compared with lead deposited on soil. Thus, lead in dust can be high and can accumulate in the interiors of dwellings, as well as in the outside surroundings. Two features of dust are important: concentration and amount. The concentration of lead in street dust may be the same in rural and urban environments, but the amount of dust may differ by a wide margin. Second, each category represents a different combination of sources. Household dusts contain some atmospheric lead, some lead paint, and some soil lead. Street dusts contain atmospheric, soil, and occasionally paint lead.[5]

Soil and dust contains lead that has settled from leaded paint, gasoline, and stationary sources; therefore, exposure estimates can only be roughly calculated as the sum of these three categories.[6] Soil and dust lead is the primary long-term repository for lead exposures; this pathway is a major contributor to overall lead exposure because of the hand-to-mouth activity of children.

In considering the impact of street dust on the human environment, the obvious question arises as to whether lead in street dust varies with traffic density. In a transect through Minneapolis and St. Paul, Mielke and associates found soil lead concentrations 10 - to - 1,000-fold higher near major interstate highways.[7] Warren and colleagues reported 20,000 μg Pb/g street dust in a heavily trafficked area.[8]

A number of studies have been conducted in urban areas to determine the effect of soil and dust contamination on the lead levels of children.[9-16] The data indicate that as blood lead increased from 10.4 to 16.6 μg/dl, surface soil scraping increased from 0 to 1,000 ppm.

A Danish study found that there is an association between individual traffic exposure and individual lead absorption.[17] An association between high lead and high traffic density was found in children at the age of 6 months to 2 years. This was a dose-response relationship and was not accounted for by other possible major sources, such as household water or parental occupation.

Dust is also a normal component of the home environment. It accumulates on all exposed surfaces, especially furniture, rugs, and windowsills. Although homemakers may take great care to remove this dust, these measures are usually inadequate to eliminate exposure to lead dust. Some households do not practice regular dust removal, and dust may be brought into the house from outside in amounts too small for efficient removal but containing lead concentrations much higher than normal baseline values.

In Omaha, Nebraska, Angle and McIntire found that lead in household dust ranged from 18 to 5,600 μg/g.[18] Clark and associates found household dusts in Cincinnati ranged from 70 to 16,000 μg Pb/g, but much of the variations could be attributed to housing quality.[19] Public housing averaged 350 μg Pb/g, rehabilitated 600 μg Pb/g, and averages in private housing ranged from 1,400 to 3,000 μg Pb/g, based on external estimates of condition, from satisfactory to deteriorating to dilapidated. In Lancaster, England, a region of low industrial lead emissions, Harrison found that household dust ranged from 510 to 970 μg Pb/g, with a mean of 720 μg Pb/g.[20] They observed that dust contained soil particles, carpet and clothing fibers, animal and human hairs, food particles, and an occasional chip of paint. Studies have found a direct correlation between atmospheric lead concentrations and the concentration of lead found in household dust.[21]

Urban atmospheres have more airborne lead than do nonurban atmospheres; therefore, there are increased amounts of lead in urban household and street dust. Reported means of urban dusts range from 500 to 3,000 μg Pb/g.[5]

Food is also a source of lead exposure. In fact, it is the most significant route of exposure for adults and older children who do not ingest lead-contaminated dust. Atmospheric lead may be added to food crops in the field or pasture, during transportation to the market, during processing, and during kitchen preparation. Metallic lead, mainly solder, may be added during processing and packaging. Other sources of lead, as yet undetermined, increase the lead content of food between the field and dinner table. Processing of foods, particularly canning, can significantly add to their background lead content, although this is lessening as the canning industry moves away from lead solder. Home food preparation can also be a source of additional lead where food preparation surfaces are exposed to moderate amounts of lead in household dust. American children, it is estimated, consume 21 μg Pb/day in food and beverages.[5]

Lead is also found in drinking water, either from contamination of the water source or from the use of lead materials in the water distribution system. Approximately 15% of the lead consumed in drinking water and beverages is of direct atmospheric origin; 60% comes from solder and other metals.[5] The EPA recently lowered the allowable level of lead in residential water to 15 ppb. They estimate that drinking water represents 20% of a child's total exposure to lead.

Smoking is another source of lead exposure in adults. Although there are no solid data, it is possible that children who are exposed to passive cigarette smoke also are exposed to lead in that smoke. One study has indicated that smoking 30 cigarettes per day results in lead intake equivalent to that of inhaling lead in the ambient air at a level of 1.0 $\mu g/m^3$.[5]

HOW MUCH LEAD IS TOO MUCH LEAD?

When we move from observations and measurements to interpretations of clinical significance, we begin to lose the scientific foundation. This is true with regard to many of the questions being asked by an anxious public and its regulatory agencies. One question without a clear-cut answer is: How much lead in various environmental sources, particularly dust and dirt, is too much? At what level does the dust or dirt become a clear-cut hazard? To answer these questions one would need to know two critical facts, both of which are uncertain. The first is how those levels translate to levels in the bodies of exposed children. The second (to be discussed below) is how levels in the bodies of children translate to adverse health effects. The fact that direct translations from levels in environmental sources to adverse health effects cannot be made has led to conflicting regulations and information for anxious parents.

Even the levels set by the various government agencies are ambiguous. The EPA states that a level over 500 ppm lead in soil qualifies a front yard as a Superfund clean-up site. By this criterion, much of our urban soil qualifies. Individual testing laboratories will tell anxious parents that 500 ppm is "too high for comfort." When asked, "What is a 'safe' level?" these same laboratories will answer that there is no definitive answer.

The EPA has said that household window wells should have levels of lead in the dust less than 800 μg Pb/cm^2. The dust in the windowsill should have levels less than 500 μg Pb/cm^2 and the floor less than 200 μg/cm^2. It is unknown, however, just what these numbers mean relative to the exposure of any given child. If the windows are never opened and the child does not eat from the floor, then these levels have little meaning. If, on the other hand, a child—through play and hand-to-mouth contact—is ingesting a fair amount of this dust, these levels are too high. The point is, there is no reasonable way, based upon existing data, to set such levels because the issue of lead exposure is so complex.

This is not to say that regulations shouldn't be set. It just means that for any given child, such policies have no clear meaning. If the dust in a house is within these limits and the soil in the front yard is over 500 ppm of lead and the water has a lead level of 15 ppb, what does this mean to the average preschooler?

WHAT DO WE KNOW VERSUS WHAT IS BEING CLAIMED ABOUT THE NEUROLOGIC EFFECTS OF LEAD EXPOSURE IN CHILDREN?

Why all the media attention and panic today? What happened to make millions of parents take notice and, for some, spend thousands or tens of thousands of dollars getting rid of the lead in their homes and subjecting their children to numerous tests and treatments, some of which have undesirable side effects?

What happened was the research of Dr. Herbert Needleman, Dr. David Bellinger, and several other investigators who said that low levels of lead, much lower than previously thought, can cause neurological, cognitive, and behavioral damage to our children, especially our pre-school children. But, before looking at these new claims about the toxicity of lead, let's look at the data about which we are relatively certain.

Lead exposure at higher levels has been virtually ignored in the most recent literature, but it is what we know best. What has happened is that the effects of lead at high levels have been ascribed, without clear scientific support, to lead exposure at low levels.

In general, lead levels of 40 to 70 μg/dl have historically been considered moderate lead poisoning. When levels are above 100, the risk of encephalopathy (brain dysfunction) is great but unpredictable. The onset of clinical symptoms of acute high-level lead poisoning in the young child usually is abrupt, with the appearance over 1 to 5 days of persistent and forceful vomiting, ataxic gait, seizures, alterations in the state of consciousness, and, finally, coma and death. These may be preceded by several weeks of irritability and decreased play activity.

As in adults, lead affects many organ systems in children. The most serious of these effects is the severe, irreversible central nervous system damage manifested in the signs and symptoms of brain damage. In children, effective blood lead levels for producing brain damage or death are lower than for adults, starting at approximately 80 to 100 μg/dl. Permanent severe mental retardation and other marked neurological deficits are among lasting neurological sequelae typically seen in cases of nonfatal childhood lead poisoning. Other overt neurological signs and symptoms of lead intoxication that do not lead to brain damage are evident in children at lower blood levels. Peripheral neuropathies have been detected in some children with blood lead levels as low as 40 to 60 μg/dl. Chronic kidney damage is most evident at high exposure levels (over 100 μg/dl), but may also exist at levels of 70 to 80 μg/dl. In addition, colic and other overt gastrointestinal symptoms clearly occur at similar or still lower blood lead levels in children, at least down to 60 μg/dl. Frank anemia is also evident by 70 μg/dl. Damage to the hematological system may begin as low as 40 μg/dl. All of these effects are reflective of the widespread, marked impact of lead on the normal physiological functioning of many different organ systems; some are evident in children at blood lead levels as low as 40 μg/dl. All of them are widely accepted as being clearly adverse health effects both by regulatory agencies and by physicians and scientists generally.[5]

At higher levels, lead has rather clearly defined effects on hemoglobin concentration in blood and P-450 activity in children. These effects are not likely to be manifested at levels of lead exposure below approximately 30 to 40 μg/dl. Thus, in estimating the thresholds for hematological effects, it has been possible to rely upon objective, quantifiable measurements.

For these reasons, the hematopoietic effects of lead have historically been used by regulatory bodies for estimating acceptable limits of human exposure.[22] Recently, however, the CDC altered these criteria and changed the acceptable limits, using the research of several physicians, most notably that of Dr. Needleman, on the alleged neuro-cognitive effects of very low level lead exposure.

When discussing the neurocognitive effects of lead, the literature describes effects that occur *below certain levels*, rather than at or above a specific level. Thus, the reports upon which the CDC relied to change the standard suggest that there are neurological and cognitive effects at levels *below 25 μg/dl.* Table 15.2 illustrates the thinking used by the CDC regarding the levels at which one might expect to see various health effects. The effects at the lowest levels are opined, but not clearly established.

COMMONLY CITED FINDINGS REGARDING LOW-LEVEL LEAD AND INTELLECTUAL FUNCTION

One of the first studies to attempt to deal seriously with the question of low-level lead's health effects was that of Needleman and associates.[24] The authors drew subjects from a population of over 3,000 first and second graders in two Massachusetts communities. Lead was measured in shed deciduous teeth and 100 children were classified as having "low" lead exposure (tooth lead level (PbT) < 10 ppm) or "high" (PbT > 20 ppm). The authors found 39 non-lead covariables that could confound their results; they were able to control for four of these: maternal age, gravidity, social class, and parental intelligence. They found that the high PbT group performed "significantly" less well on the full scale and verbal subscale of the Wechsler Intelligence Scale for Children-Revised (WISC-R). The difference was approximately 4.5 IQ points.

Ernhart and coauthors have also found that as lead exposure goes up, intelligence seems to go down.[25] They followed a sample of preschool children in New York City who had PbB levels in the range of 40 to 70 μg/dl and a comparison sample of children with lower PbB levels.[26] In the earlier study, preschool PbB levels in the range of 40 to 70 μg/dl were significantly and inversely related to scores on the McCarthy Scales of Children's Abilities (MSCA), a standardized measure of intellectual ability and achievement for preschool and school-age children. Upon follow-up, school-age PbB levels were lower than preschool levels, but not uniformly low by current clinical standards (Low Group Mean $= 21.3 \pm 3.7$ μg/dl, High Group M $= 32.4 \pm 5.3$ μg/dl). After controlling for the child's sex and parental intelligence, there were significant and inverse associations between school-age PbB levels and the general cognitive index and verbal and motor subscales.

Table 15.2 Adverse Effects of Lead

Neurologic effects	Hemesynthesis effects	Other effects	Lowest PbB (μg/dl) effect seen
Deficits in neurobehavioral development (Bayley and McCarthy Scales); electrophysiological changes	ALA-D inhibition	Reduced gestational age and weight at birth; reduced size up to 7-8 years of age	10-15 (prenatal and postnatal)
	EP elevation	Impaired vitamin D metabolism; Py-5-N inhibition	15-20
Lower IQ, slower reaction time (studied cross-sectionally)			<25
Slowed nerve conduction velocity			30
	Reduced hemoglobin; elevated CP and ALA-U		40
Peripheral neuropathies	Frank anemia		70
Encephalopathy		Colic and other GI effects; kidney effects	80-100

Source: Centers for Disease Control, Childhood lead poisoning - United States: report to the Congress by the Agency for Toxic Substances and Disease Registry, *MMWR*, 37, 481, 1988.

Parents looking at these numbers naturally are terrified, especially if their child has a lead level between 10 and 25 μg/dl. After all, the chart says that their child will have a lower IQ and deficits in neurobehavioral development. What the average parent does not know, however, is the extreme controversy surrounding these numbers and the research from which they were derived. To elucidate, let us examine what we do and do not know about the neurocognitive effects of lead exposure.

In studying lower levels of lead exposure, however, Ernhart and colleagues found that the correlations between low-level lead exposure and intelligence test scores were attenuated, not statistically significant, and not consistent in direction when relevant confounding variables were considered.[27] The extent of effects of both prenatal and preschool lead exposure to intellectual development were small, not statistically significant, and not consistent in direction. Hammond and Deitrich confirm that Ernhart and co-authors did not interpret their findings as positive. Factors that affected interpretation of the data included multiple statistical tests and detailed controls over covariates.[28]

Much of the cross-sectional lead research that followed the original Needleman and Ernhart work were attempts to replicate or expand the efforts of these two researchers. Much of this work was carried out by European researchers.[29-35] Overall, the results of these studies suggest that lead exposure capable of resulting in PbB levels of 25 μg/dl and above are likely to produce deficits in intellectual performance. Others, however, have yielded somewhat conflicting results.[36-38] The deficits are small until levels have reached at least 40 μg/dl and are not clearly present below 25 μg/dl.

One of the largest, most recent studies is the cross-sectional study conducted in Edinburgh, Scotland. Fulton et al. studied 501 children aged 6 to 9 years who were at risk for lead exposure because of a plumbosolvent water supply and a large number of houses with lead plumbing.[39] The British Ability Scales (BASC), a test of cognitive ability and academic achievement, were administered to all children. Lead was assessed in blood near the time of testing and averaged 11.5 μg/dl. Only 1.9% had PbB levels above 25 μg/dl. Thirty-three confounding variables were selected based on both theoretical and empirical criteria and included as potential confounders in multiple regression analyses. The data suggested statistically significant relationships between PbB level and total BASC score and the quantitative and reading subscales. The data also suggested a clear dose-response relationship with no evidence of a threshold. The differences, however, were small; between the lowest and highest PbB groups, there was a difference of only 5.8 points.

Many other recent studies of low to moderate lead exposure appear to confirm the earlier findings of Needleman.[40-45] However, it is clear that at quite low levels of exposure, the effects may be so small that a very large sample is required to detect it, and the accuracy of these tests is such that 4 or 5 IQ points may not be a significant difference.[28] In addition, there has been no rigorous controlling for all of the confounding variables.

What Are the Criticisms of This Work?

It should be noted that there are many difficulties in studying the central nervous system effects of lead exposure, many of which have been encountered and inadequately dealt with by Dr. Needleman and others investigating the effects of lead. All of the profound problems of experimental design (Chapters 3 and 5) plague these studies. Measurement of the independent variable, the lead level, can be a problem. Controlling for confounders, covariates, and biases present substantial and even irreconcilable difficulties. Most critical, perhaps, is the measurement of the dependent variable: subtle neurobehavioral effects such as classroom behavior and IQ, soft end points with limited sensitivity or specificity.

Lead Levels: The Independent Variable

Insults to the developing brain may or may not be reversible. Consequently, it is difficult to determine whether any measured insult is due to current or past exposure, at a time when the child had a very high but undetermined lead level.

Investigators usually must assume that the current lead level adequately measures the entire exposure history of the child up until the time that he or she was tested. In fact, the investigators generally have very little information regarding the true exposure history of the children being studied. The children may have been much more exposed for a much longer period of time than the investigators knew at the time of the study. There is no way of knowing, in most cases, whether the one or two blood lead levels represent an exposure history that has really prevailed long enough to affect the child. In addition, there is usually no information regarding the age level at which the child was exposed, which has profound consequences upon assessing the ability of the lead to affect the developing brain. A "high" lead level at one point in time may actually reflect very recent short-term exposure that would be unlikely to have caused the abnormalities found on testing. Without knowing this, the researcher is not prompted to look for other possible causes. In short, investigators are usually unable to document the lead exposure history in such a way that we know there is a reasonable separation of exposure levels between the experimental and the control groups. Consequently, investigators have great difficulty answering critical questions regarding whether the timing of the exposure is important and whether there is a dose-duration effect on later development.

To illustrate the complexity of determining exposure using blood lead levels, consider the long half-life of the substance. When a single tracer dose of lead is given, the blood lead half-life is on the order of 18 days.[28] Under the more realistic conditions of exposure, that is, continuous small amounts of lead over months or years, the half-life is considerably longer.

It should be kept in mind that when looking at blood lead levels in this range, there is variability in the test itself, in any test, which makes it more difficult to say that a level of 12 μg/dl confers greater risk than a level of 8 μg/dl. The test result of 12 μg/dl may *actually be* a level of 8 μg/dl. The differences between any two test results at these levels is quite muddy.

Confounders, Covariates, and Biases

If you recall in Chapters 3 and 5, we discussed experimental design and the interpretation of data. The scientific exploration of the effects of low-level lead is prone to the influences of confounders, covariates, and biases.

A major problem derives from the end points or factors being investigated. These are not gross or readily objectifiable neurological dysfunction. Rather, they are subtle effects detectable through the subjective reporting of observers such as teachers or through error-prone neuropsychological testing.

Even if such behavioral or cognitive alterations actually exist, it is difficult to distinguish between neurobehavioral effects of lead and effects due to the many social and medical factors that are known to have important impacts upon neurobehavioral development.[45] Many unmeasurable and uncontrollable factors can produce subtle effects in a child's ability to learn and to concentrate. Even when experimental and control groups are fairly homogeneous on correlated variables such as race, socioeconomic status, age, and educational level, there are other, less concrete variables that may have profound influence: the family's rearing style; cultural opportunities provided within the family or by the preschool or day-care providers; variations among quality of preschool and/or day-care situations; familial attitudes toward education; social values of the nuclear and extended family; parental personalities and attitudes toward their children; undiscovered family pathology, such as alcoholism; undisclosed mental illness in one or both parents; undiscovered emotional or physical abuse; and, finally, individual personality differences, completely separate from lead exposure, of individual children, such as shyness, unwillingness to assert oneself, or any number of personality differences. Even whether a child had a nightmare the night before the test, or was coming down with a cold, or any one of a hundred factors might result in a four-point difference on an IQ test and slow a reaction time by a millisecond.

This brings us to the minuscule differences today's researchers have identified. Keep in mind that this research has not revealed a group of youngsters made mentally retarded by lead levels of 15 μg/dl. Rather, the differences identified are so small that for any given child they may be meaningless.

There is also the opportunity for misapplication of statistical procedures, evident in many of the earlier studies of lead toxicity. According to Bornschein and coauthors, many studies tested their findings by utilizing 10 to 15 separate "t"-tests.[46] This procedure inflates the chance of obtaining a significant "t" value by chance alone. When you add the countless statistical manipulations, and the chances for error, distortion, and exaggeration inherent in the use of such statistics, the probability of inaccurate results is overwhelming.

The difficulty inherent in studying this area is seen clearly in the history of the research. Early studies, that is, those conducted between the 1940s and 1970s, often contained several flaws attributable to these difficulties. More adequately designed cross-sectional and prospective studies have been carried out from the 1970s to the present.[28] These newer studies have improved the assessment of lead exposure by obtaining reliable measurements of lead in shed deciduous teeth (PbT) or serial lead in blood (PbB) determinations spanning a relatively long time in the child's life. Newer studies have also attempted to recruit subjects in an unbiased manner from a general population of children not already known to possess cognitive deficits or psychiatric disorders. Further, serious attempts were made to identify, reliably assess, and statistically control for other factors that affect neurobehavioral development, many of which tend to be correlated with risk for lead exposure, for example, social class, caretaking quality, nutrition, and obstetrical and postnatal complications.[28]

Still, many of the difficulties noted above are impossible to control, and there is no clear evidence that lead exposure at low levels causes the type of damage that has set off the explosion of media coverage of this topic and its associated regulatory fervor.

Perhaps one of the most damning criticisms of Dr. Needleman's work, and that of other researchers who have reached similar conclusions, is that numerous similar tests must be done to find one or two abnormalities. In other words, critics charge that researchers had to perform 10 tests of memory to find deficits in 1 of the tests. It is charged that the test revealing an abnormality is then the only test used in the final analysis. This analytical or observer bias is commonly found in the neurobehavioral lead literature. Critics also note that sample sizes had to be very large in order to find abnormalities on any tests. In any given group of, say, 50 or 100 children, it was not possible to find any statistically significant "abnormalities" among children who had lead levels of 10 to 25 μg/dl versus those who had lead levels below 10 μg/dl.

In some studies, teachers' perceptions of children's behavior were used to determine lead effects. Such measures were used to state that children with blood-lead elevations had higher incidences of classroom behavior and learning problems. The EPA's 1983 Expert Committee on Pediatric Neurobehavioral Evaluations, however, concluded that "...the relationship between dentine lead levels and teachers' ratings of classroom behavior cannot be safely attributed to the effects of lead...."[47] They also found several other errors in analysis in the most recent lead studies, confirming the difficulty inherent in research in this area. In fact, in reviewing both Dr. Ernhart's and Dr. Needleman's work, the committee found that they neither could confirm nor refute the hypothesis that low-level lead exposure in children leads to neuropsychological deficits.

What Do the "Abnormalities" Cited in These Studies Really Mean?

Even if we were to overcome the methodological impediments and interpretations noted above, another question remains. The subtle changes in cognitive testing noted by those who led the antilead charge, particularly Dr. Needleman, are quite small indeed: four IQ points, as an example. A critical question is: What does that mean? First of all, IQ testing in children is a process subject to quite a substantial error rate and significant variability. Thus, it is not clear whether an IQ point difference of four might really be one, or none at all. Furthermore, what if an IQ actually were changed by four points, how would this affect an individual's life? What, if any role would it play in an individual's success or failure in life? This remains, as far as I know, an unanswered question. None of this is to argue that lead poisoning is okay or that we should readily accept IQ reductions (if true), but it does raise the question of the significance of the alleged findings.

Another point is: What does all of this have to do with levels of 15 or 20 μg/dl. Needleman's studies and essentially all others explored differences in performance and cognitive function in children whose lead levels were considerably higher than this and compared them with low-exposure controls. Dr. Ernhart's early positive studies found alterations in children with levels from 40 to 70 μg/dl and compared them with the "low" group at 21. Even Dr. Needleman used as his study group children with levels above 20 μg/dl.

This brings us to the question of popular perception and regulatory responses. There is a common fear and belief that children are being affected severely by even modest amounts of lead. By adopting a 10μg/dl standard, the CDC has placed Americans on notice that there is a widespread threat to children's health by the lead menace. Twenty percent or more of our children are likely involved, using these extremely conservative criteria. The data connecting such levels to adverse health effects are essentially nonexistent and what data do exist do not confirm these dire estimates. People are worried, and I fear that our regulatory machinery has, in this case, contributed to false perceptions and intensified unnecessarily those fears.

REFERENCES

1. Waldman, S., Lead and your kids, *Newsweek*, July 15, 1991.

2. Mushak, P. and Crocetti, A.F., Determination of numbers of lead-exposed American children as a function of lead source: integrated summary of a report to the U.S. Congress on childhood lead poisoning, *Environ. Res.*, 50, 673, 1989.

3. Kazantzis, G., Lead: sources, exposure and possible carcinogenicity, *IARC Sci. Publ.*, 71, 103, 1986.

4. Morbidity and Mortality Weekly Report (MMWR), Centers for Disease Control, Atlanta, Reprinted in *JAMA*, 260, 1523, 1988.

5. U.S. Environmental Protection Agency, Health Effects Research Lab, Criteria and Special Studies Office, *Air Quality Criteria for Lead*, National Technical Information Service, Springfield, VA, 1986.

6. Agency for Toxic Substances and Disease Registry, *The Nature and Extent of Lead Poisoning in Children in the United States: A Report to Congress*, U.S. Department of Health and Human Services, Public Health Services, Atlanta, 1988.

7. Mielke, H., Blake, B., Burroughs, S. and Hassinger, N., Urban lead levels in Minneapolis: the case of the Hmong children, *Environ. Res.*, 34, 64, 1984.

8. Warren, H.V., Dlavault, R.E., Fletcher, K. and Wilks, E., Variations in the copper, zinc, lead, and molybdenum content of some British Columbia vegetables, in *Trace Substances in Environmental Health-VI (Proceedings of the University of Missouri's 4th Annual Conference on Trace Substances in Environmental Health)*, Hemphill, D.D., Ed., University of Missouri, Columbia, 1971, 94.

9. Galke, W.A., Hammer, D.I., Keil, J.E. and Lawrence, S.W., Environmental determinants of lead burdens in children, *International Conference on Heavy Metals in the Environment*, Institute for Environmental Studies, Toronto, 1975, 53.

10. Stark, A.D., Quah, R.F., Meigs, J.W. and DeLouise, E.R., The relationship of environmental lead to blood-lead levels in children, *Environ. Res.*, 27, 372, 1982.

11. Bornschein, R.L., Succop, P., Draft, K.M., Clark, C.S., Peace, B. and Hammond, P.B., Exterior surface dust lead, interior house dust lead, and childhood lead exposure in an urban environment, in *Trace Substances in Environmental Health II, A Symposium*, University of Missouri, Columbia, 1986.

12. Reeves, R., Kjellstrom, T., Dallow, M. and Mullins, P., Analysis of lead in blood, paint, soil, and house dust for the assessment of human lead exposure in Aukland, *N. Z. J. Sci.*, 25, 221, 1982.

13. Rabinowitz, M., Leviton, A., Needleman, H., Bellinger, D. and Waternauz, C., Environmental correlates of infant blood lead levels in Boston, *Environ. Res.*, 38, 96, 1985.

14. Mielke, H.W., Anderson, J.C., Berry, K.J., et al., Lead concentrations in inner city soils as a factor in the child lead problem, *Am. J. Public Health*, 73, 1366, 1983.

15. Elwood, P.C., The sources of lead in blood: a critical review, *Sci. Total Environ.*, 52, 1, 1986.

16. Duggan, M.J. and Inskip, M.J., Childhood exposure to lead in surface dust and soil: a community health problem, *Public Health Rev.*, 13, 1, 1985.

17. Lyngbye, T., Hansen, O.N., Grandjean, P., Trillingsgaard, A. and Beese, I., Traffic as a source of lead exposure in childhood, *Sci. Total Environ.*, 71, 461, 1988.

18. Angle, C.R. and McIntire, M.S., Environmental lead and children: the Omaha study, *J. Toxicol. Environ. Health*, 5, 855, 1979.

19. Clark, C.S., Bornschein, R.L., Succop, P., Que Hee, S.S., Hammond, P.B. and Peace, B., Condition and type of housing as an indicator of potential environmental lead exposure and pediatric blood lead levels, *Environ. Res.*, 38, 46, 1985.

20. Harrison, R.M., Toxic metals in street and household dusts, *Sci. Total Environ.*, 11, 89, 1979.

21. Davies, B.E., Elwood, P.C., Gallacher, J. and Ginnerver, R.C., The relationships between heavy metals in garden soils and house dusts in an old lead mining area of North Wales, Great Britain, *Environ. Pollut. [B]*, 9, 255, 1985.

22. Centers for Disease Control, Preventing Lead Poisoning in Young Children: A Statement by the Centers for Disease Control, Centers for Disease Control, Atlanta, 1985.

23. Centers for Disease Control, Childhood lead poisoning - United States: report to the Congress by the Agency for Toxic Substances and Disease Registry, *MMWR*, 37, 481, 1988.

24. Needleman, H.L., Gunnoe, C., Leviton, A., Reed, R., Peresie, H., Mahere, C. and Barrett, P., Deficits in psychologic and classroom performance of children with elevated dentine levels, *N. Engl. J. Med.*, 300, 680, 1979.

25. Ernhart, C.B., Landa, B. and Schell, N.B., Subclinical levels of lead and developmental deficit: a multivariate follow-up reassessment, *Pediatrics*, 67, 911, 1981.

26. Perino, J. and Ernhart, C.B., The relation of subclinical lead level to cognitive and sensorimetric impairment in black preschoolers, *J. Learn. Disabil.*, 7, 26, 1974.

27. Ernhart, C.B., Morrow-Tlucak, M., Wolf, A.W., Super, D. and Drotar, D., Low level lead exposure in the prenatal and early preschool periods: intelligence prior to school entry, *Neurotoxicol. Teratol.*, 11, 161, 1989.

28. Hammond, P.B. and Dietrich, K.N., Lead exposure in early life: health consequences, *Rev. Environ. Contam. Toxicol.*, 115, 91, 1990.

29. Hatzakis, A., Kokkevi, A., Katsouyanni, K., Maravelias, K., Salaminios, R., Kalandidi, A., Koutselinis, A., Stefanis, K. and Trichopoulos, D., Psychometric intelligence and attentional performance deficits in lead-exposed children, in *Heavy Metals in the Environment*, Volume 1, Lindberg, S.E. and Hutchinson, T.C., Eds., CEP Consultants Ltd., Edinburgh, 1987, 204-209.

30. Yule, W., Lansdown, R., Millar, R. and Urbanowicz, M.A., The relationship between blood lead concentration, intelligence and attainment in a school population: a pilot study, *Dev. Med. Child Neurol.*, 23, 567, 1981.

31. Annest, J.L., Pirkle, J.L., Makuc, D., Neese, J.W., Bayse, D.D. and Kovar, M.G., Chronological trend in blood lead levels between 1976 and 1980, *N. Engl. J. Med.*, 308, 1373, 1983.

32. Smith, M., Recent work on low level lead exposure and its impact on behavior, intelligence, and learning: a review, *J. Am. Acad. Child Psych.*, 24, 24, 1985.

33. Medical Research Council, *The Neuropsychological Effects of Lead in Children. A Review of Recent Research 1979-1983*, Medical Research Council, London, 1984, 64.

34. Winneke, G., Neurobehavioral and neuropsychological effects of lead, in *Lead Versus Health*, Rutter, R.M. and Jones, R.R., Eds., John Wiley & Sons Ltd., England, 1983, 249.

35. Winneke, G., Kraemer, U., Brockhaus, A., Ewers, U., Kujanek, G., Lechner, H. and Janke, W., Neuropsychological studies in children with elevated tooth-lead concentrations, *Int. Arch. Occup. Environ. Health*, 51, 231, 1983.

36. Harvey, P.G., Lead and children's health: recent research and future questions, *J. Child Psychol. Psychiat.*, 25, 517, 1984.

37. Yule, W. and Lansdown, R.G., Blood lead concentrations in school age children, intelligence, attainment, and behavior. Paper presented at the annual conference of the British Psychological Society, York, April 1983.

38. Pocock, S.J., Ashby, D. and Smith, M.A., Lead exposure and children's intellectual performance, *Int. J. Epidem.*, 16, 57, 1987.

39. Fulton, M., Raab, G., Thomson, G., Laxen, D., Hunter, R. and Hepburn, W., Influence of blood lead upon the ability and attainment of children in Edinburgh, *Lancet*, 1, 1221, 1987.

40. Hansen, O.N., Trillingsgaard, A., Beese, I., Lyngbye, T. and Grandjean, P., A neuropsychological study of children with elevated dentine lead level, in *Heavy Metals in the Environment*, Volume 1, Lindberg, S.E. and Hutchinson, T.C., Eds., CEP Consultants Ltd., Edinburgh, 1987, 49-53.

41. Lyngbye, T., Hansen, O.N. and Grandjean, P., Neurological deficits in children: medical risk factors and lead exposure, *Neurotoxicol. Teratol.*, 10, 531, 1989.

42. Ferguson, D.M., Ferguson, J.E., Horwood, L.J. and Kinzett, N.G., A longitudinal study of dentine lead levels, intelligence, school performance and behavior. Part I. Dentine levels and exposure to environmental risk factors, *J. Child Psychol. Psychiatry*, 29, 781, 1988.

43. Ferguson, D.M., Ferguson, J.E., Horwood, L.J. and Kinzett, N.G., A longitudinal study of dentine lead levels, intelligence, school performance and behavior. Part II. Dentine level and cognitive ability, *J. Child Psychol. Psychiatry*, 29, 793, 1988.

44. Ferguson, D.M., Ferguson, J.E., Horwood, L.J. and Kinzett, N.G., A longitudinal study of dentine lead levels, intelligence, school performance and behavior. Part III. Dentine levels and attention/activity, *J. Child Psychol. Psychiatry*, 29, 811, 1988.

45. Schroeder, S.R., Hawk, B., Otto, D.A., Mushak, P. and Hicks, R.E., Separating the effects of lead and social factors on IQ, *Environ. Res.*, 38, 144, 1985.

46. Bornschein, R., Pearson, D. and Reiter, L., Behavioral effects of moderate lead exposure in children and animal studies: part 1, *CRC Crit. Rev. Toxicol.*, 8, 43, 1980.

47. U.S. Environmental Protection Agency, Independent Peer Review of Selected Studies Concerning Neurobehavioral Effects of Lead Exposures in Nominally Asymptomatic Children: Official Report of Findings of an Interdisciplinary Expert Review Committee, U.S. Environmental Protection Agency, Washington, DC, 1983.

GENERAL REFERENCES

Baloh, R., Sturm, R., Green, B. and Gleser, J., Neuropsychological effects of chronic asymptomatic increased lead absorption, *Arch. Neurol.*, 32, 326, 1975.

Beattie, A.D., Moore, M.R., Goldberg, A., Finlayson, M.J., Graham, J.F., Mackie, E.M., Main, J.C., McLaren, D.A., Murdoch, K.M. and Steward, G.T., Role of chronic low-level lead exposure in the etiology of mental retardation, *Lancet*, 1, 589, 1975.

Bellinger, D., Leviton, A. and Sloman, J., Antecedents and correlates of improved cognitive performance in children exposed in utero to levels of lead, *Environ. Health Perspect.*, 89, 5, 1990.

Bellinger, D., Leviton, A., Waternaux, C., Needleman, H. and Rabinowitz, M., Longitudinal analyses of prenatal and postnatal lead exposure and early cognitive development, *N. Engl. J. Med.*, 316, 1037, 1987.

David, O.J., Clark, J. and Voeller, K., Lead and hyperactivity, *Lancet*, 2, 900, 1972.

David, O.J., Grad, G., McGann, B. and Koltun, A., Mental retardation and "non-toxic" lead levels, *Am. J. Psychiatry*, 139, 806, 1982.

De La Burde, B. and Choate, M.S., Early asymptomatic lead exposure and development at school age, *J. Pediatr.*, 82, 638, 1975.

Dietrich, K.M., Succop, P.A., Bornschein, R.L., Kraffte, K.M., Berger, O., Hammond, P.B. and Buncher, C.R., Lead exposure and neurobehavioral development in later infancy, *Environ. Health Perspect.*, 89, 13, 1990.

Ernhart, C.B., Lead levels and confounding variable, *Am. J. Psychiatry*, 139, 1524, 1982.

Harvey, P.G., Lead and children's health: recent research and future questions, *J. Child Psychol. Psychiatry*, 25, 517, 1984.

Harvey, P.G., Hamlin, M.N., Kumar, R. and Delves, H.T., Blood lead, behavior and intelligence test performance in preschool children, *Sci. Total Environ.*, 40, 45, 1984.

Hebel, J., Krisch, D. and Armstrong, E., Mental capability of children exposed to lead pollution, *Br. J. Prevent. Soc. Med.*, 30, 170, 1976.

Kotok, D., Kotok, R. and Heriot, J.T., Cognitive evaluation of children with elevated blood lead levels, *Am. J. Dis. Child*, 131, 791, 1977.

Landrigan, P.J., Whitworth, R.H., Baloh, R.W., Staehling, N.W., Barthel, W.F. and Rosenblum, B.F., Neuropsychological dysfunction in children with chronic low level lead absorption, *Lancet*, 1, 708, 1975.

Landsdown, R., Shepherd, J., Clayton, B., Delves, H.I., Graham, P.J. and Turner, W.G., Blood levels, behavior and intelligence: a population study, *Lancet*, 1, 538, 1974.

Landsdown, R., Yule, W., Urbanowicz, M. and Hunter, J., The relationship between blood-level concentrations, intelligence, attainment and behavior in a school population, *Int. Arch. Environ. Health*, 57, 225, 1986.

Marshall, E., EPA faults classic lead study, *Science*, 222, 906, 1983.

McBride, W., Black, B. and English, B., Blood lead levels and behavior of 400 preschool children, *Med. J. Aust.*, 2, 26, 1982.

Moore, M.R., Meredith, P.A. and Goldberg, A., A retrospective analysis of blood-lead in mentally retarded children, *Lancet*, 1, 717 1977.

Needleman, H.L., Low level lead exposure in the fetus and young child, *Neurotoxicology*, 8, 389, 1987.

Needleman, H.L., The persistent threat of lead: a singular opportunity, *Am. J. Public Health*, 79, 643, 1989.

Needleman, H.L., What can the study of lead teach us about other toxicants? *Environ. Health Perspect.*, 86, 184, 1990.

Needleman, H.J., The future challenge of lead toxicity, *Environ. Health Perspect.*, 89, 85, 1990.

Needleman, H.L., Davidson, I., Sewell, E. and Shapiro, I.M., Subclinical lead exposures in Philadelphia school children: identification by lead analysis, *N. Engl. J. Med.*, 290, 245, 1974.

Needleman, H.L., Leviton, A. and Bellinger, D., Lead associated intellectual deficit, *N. Engl. J. Med.*, 306, 367, 1982.

Needleman, H.L., Schell, A., Bellinger, D., Leviton, A. and Alfred, E.N., The long-term effects of exposure to low levels of lead in childhood: an 11-year follow-up report, *N. Engl. J. Med.*, 322, 83, 1990.

Paterson, L.J., Raab, G.M., Hunter, R., Laxen, D.P., Fulton, M., Fell, G.S., Halls, D.J. and Sutcliffe, P., Factors influencing lead concentrations in shed deciduous teeth, *Sci. Total Environ.*, 74, 219, 1988.

Ratcliffe, J., Developmental and behavioral functions in young children with elevated blood lead levels, *Br. J. Prevent. Soc. Med.*, 31, 258, 1977.

Rutter, M., Raised lead levels and impaired cognitive/behavioral functioning: a review of the evidence, *Dev. Med. Child Neurol. [Suppl.]*, 22, 1, 1980.

Sachs, H.K., Krall, V., McCaughran, D.A., Rosenfeld, I., Yongsmith, N., Growe, G., Lazar, B.S., Novar, L., O'Connell, L. and Rayson, B., IQ following treatment of lead poisoning: a patient sibling comparison, *J. Pediatr.*, 93, 428, 1978.

Smith, M., Delves, T., Lansdown, R., Clayton, B. and Graham, P., The effects of lead exposure on urban children, *Dev. Med. Child Neurol. [Suppl.]*, 47, 25, 1983.

Thatcher, R., Lester., M., McAlaster, R. and Horst, R., Effects of low levels of cadmium and lead on cognitive functioning in children, *Arch. Environ. Health*, 37, 159, 1982.

16
AGRICULTURAL CHEMICALS AND CANCER: THE CASE OF ALAR

The more personal the threat, the greater the fear. Contaminated food is more frightening than is a contaminated ocean. Children at risk, particularly when that risk is cancer, engender the most profound of concerns. How understandable it is, therefore, that pesticides have been accorded a special place in the annals of popular worries and toxic perceptions. Residual levels of agricultural chemicals found in fruits, vegetables, and the meat of farm animals are the archetypical chemical threat, and vulnerable children are viewed as their most shocking of victims. The growing interest in "organic foods" is one apparent response to this concern. Another such popular response was Proposition 65, the California law that requires hazard warnings on fruits and vegetables in supermarkets. Emotionalism is the antithesis of rationalism. In the case of pesticides, the intensity of the emotionalism has created an enormous gulf between their perceived hazards and their established risks.

The scare over Alar exemplifies this gulf. In this contentious public debate were all of the components of science, perception, and regulation of a commercial chemical. As an illustration of popular chemical fears, the Alar story is preeminent. A public relations firm was hired by an environmental organization to manipulate public sentiment to demand a regulatory ban. The resulting distortion of science and near panic by the parents and grandparents of apple-eating children seemed to the anti-Alar forces a small price for the perceived good.

The science of this Alar saga involved limited rodent data suggesting carcinogenicity of unsymmetrical dimethylhydrazine (UDMH) a trace contaminant of Alar and a by-product that may derive from it. That data led to intense pressure from environmental organizations, particularly the Natural Resources Defense Council (NRDC), to press for a total ban of Alar. When the regulatory machinery ground too slowly, the NRDC launched an elegant, highly effective public relations campaign that translated these limited and tangential animal data into clear images of thousands of American children dying of cancer. The campaign was highly successful in carrying out its intended function: creating a perception. Its by-products included fear, distortion, mistrust, economic hardships and now lawsuits by apple growers against the NRDC.

THE ALAR HISTORY

The Alar story itself begins with the introduction of Alar in the United States for use on plants and crops. Daminozide is the accepted common chemical name. Under the trade name of Alar it was first registered by the Uniroyal Chemical Company in 1963 for use on potted chrysanthemums and in 1968 for use on apples. Since then, it has been used in the United States on various ornamental plants and on fruits, vegetables, and peanuts as a preservative and to enhance the growth and quality of the crops.

Questions concerning the carcinogenicity of daminozide began in 1973 when a study by Bela Toth of the Eppley Institute for Research showed a high incidence of tumors of the blood vessels, lung, kidney, and liver in mice exposed to the metabolite of Alar, 1,1-dimethylhydrazine.[1]

Historically, the regulatory response to these findings begins with the Delaney clause. The Delaney clause, amended to the Food, Drug and Cosmetic Act in 1958, required the FDA to reduce to zero any food additive found to be carcinogenic. It states "...no additive shall be deemed to be safe if it is found to induce cancer when ingested by man or animal, or if it is found, after tests which are appropriate for the evaluation of the safety of food additives to induce cancer in man or animal...."[2] Both the FDA and the EPA have certain jurisdictional responsibilities over the Food, Drug and Cosmetic Act and, therefore, are bound under specific instances by the Delaney clause. Traditionally, the application of the clause is applied to food additives and not to residual agricultural chemicals on foods, with one exception. When foods are processed such as, apples turned into applesauce or apple juice, the agencies have often accepted and employed the Delaney clause.

The Delaney clause was passed at a time when it seemed both eminently sensible and easy to comply with its terms. Why, after all, would we add anything to food that has been shown to cause cancer? Why would we want any such substances in our processed foods? As years progressed, the breadth of animal and cellular systems used to evaluate carcinogenicity expanded (Chapters 7, 9, and 10), and as quantitative testing became capable of finding smaller and smaller quantities of residual materials (Chapter 8), the boundaries of Delaney became broader and hazier. Chloroform, for example, a by-product of the chlorination of water, fits the Delaney requirements. It should be eliminated; yet, it is found at levels of 80 ppb or so in most municipal water supplies. Its risk is clearly considered to be outweighed by the benefits of chlorination.

Pepper, we now know, contains carcinogens; yet, it is added to food. And, innumerable valuable farms chemicals can be shown in one test or another to cause cancer in some species of animal, but we cannot do without all of them, at least, not today. At the time the act was passed, no one could envision either the extraordinarily low levels at which chemicals could be measured or the very large numbers of chemicals which would be found to satisfy certain of its criteria.

With its carcinogenicity studies Alar came into the penumbra of the Delaney clause. Based on the initial 1973 study and several later studies by Toth,[3,4] in 1980 the EPA announced plans to conduct a "special review" of Alar.[5] This process is necessary for cancellation of products if new data indicating potential carcinogenicity become available after the product is approved for use. Following private discussions between the EPA and Uniroyal, this review was temporarily put on hold, but it was reinstated in 1984 under pressure from environmental groups including the NRDC.

In 1985 the EPA published a notice in the *Federal Register* of intent to cancel registrations permitting the use of daminozide on foods.[6] The Draft Cancellation Notice was submitted for review to EPA's Scientific Advisory Panel (SAP) as required by FIFRA. The panel concluded that the bioassays used by the EPA were inadequate for quantitative risk assessment. And, under FIFRA, risk assessment would be applied. Based on the SAP recommendations, the EPA postponed cancellation actions, but required the development of better data.[7]

A variety of studies were performed between 1985 and 1989. These will be described in more detail later, but, essentially, studies were only positive for cancer when toxic levels were given. Based largely upon these positive studies, the EPA concluded that dietary exposure to UDMH represented a significant carcinogenic risk that outweighed the benefits of use of daminozide on crops and therefore the food uses would be canceled. Both UDMH and daminozide were now classified as Group B2, carcinogens. This notice of intent to cancel was published in the January 1989 *Federal Register.*[6] In that notice, the EPA extended the apple tolerance for 18 months to allow for cancellation proceedings. The reasoning behind the extension was that, according to the EPA risk assessment calculations, there would be less than one cancer per million during the 18-month period. The figure was based on the EPA estimate of a lifetime risk of 45 cancers per million exposed persons. Such a low risk was considered insufficient to warrant an emergency order or a change in the existing temporary tolerance.[7]

CREATING A PERCEPTION: THE ROLE OF THE NRDC AND THE MEDIA BLITZ

Following the EPA's proposed delay, the NRDC orchestrated a media blitz with the help of its public relations firm, Fenton Communications. Its opening salvo was aired on CBS television's *60 Minutes* on February 26, 1989, the day before the release of the NRDC publication, *Intolerable Risk: Pesticides in Our Children's Food.* On the *60 Minutes* program, Ed Bradley interviewed representatives of the NRDC who pronounced alarming statistics: Alar causes an "estimated 240 deaths per million population among children who are average consumers of Alar-treated food and a whopping 910 per 1,000,000 ('an additional cancer case for every 1,000 exposed') for heavy consumers." Children "are at increased risk because they drink 18 times as much apple juice as their mothers." "Alar is a cancer-causing agent that is used on food and the EPA knows it will cause cancer in thousands of children over their lifetime."[5]

The NRDC risk assessment was done by William Nicholson of the Mt. Sinai School of Medicine and reviewed by a scientific panel. In arriving at their frightening risk estimates, the NRDC used different numbers for its calculations than the EPA used. More important, they presented these numbers as though they were true and real, rather than the product of the theories and assumptions that underlie quantitative risk assessment (Chapter 10).

Speaking on this *60 Minutes* program, Janet Hathaway, attorney for the NRDC, stated, "Just from these eight pesticides [the NRDC reviewed, sic] over a lifetime, one child out of every 4,000 preschoolers or so will develop cancer."[5] This statement, while important for the public relations campaign of the NRDC, was misleading, inappropriately frightening, and incorrect. Even if one were to accept uncritically the risk assessment methodologies used by the NRDC to assess Alar and the other pesticides, no true scientist would suggest that the resulting numbers represented more than an assumptive possibility. To convert such mathematical guesswork into "children will develop cancer" goes beyond the data and misleads the public, only to serve a public relations mission.

Other events were staged by Fenton to heighten the fear of Alar and to intensify cries for its immediate removal. Advance interviews with major women's magazines and appearance dates with the *Donahue* show and other television programs marched forward. Most effective of all was actress Meryl Streep's announcement of the formation of a group, Mothers and Others for Pesticide Limits, at a news conference on March 1, 1989.

Lost in the one-sided presentation to the public was some of the critical contextual information. Mice responded only to very high toxic doses was one piece of that information. This enormous 23 mg/kg dose contrasted starkly with the EPA's estimation that in the 1980s the average U.S. citizen's exposure was 0.000047 mg/kg/day (or 1/600,000 of the animal dose).[8] The fact that the background intake of naturally occurring carcinogenic pesticides exceeds man-made residues by 10,000 times was another critical omission (see Chapter 8).

The Alar incident escalated. The EPA, FDA, and the Academy of Pediatrics appeared before the House of Representatives, Agriculture Committee to testify on the risks from pesticide residues in food. New York City and Los Angeles removed apples from school cafeterias. In May 1989 apple growers demanded that EPA cancel Alar registrations at once because the apple industry had lost $100 million. Legislation was introduced in Congress to carry this out. But in June, Uniroyal signed an agreement with the EPA to stop sales immediately and recall all stocks of food on which daminozide had been used. For the time being, the EPA did not need to go any further with study results or regulations.

ALAR RISK ASSESSMENTS

There were profound differences of opinion concerning carcinogenicity and safe exposure levels among the NRDC, EPA, Uniroyal, and other groups. These differences are based on the levels of the chemical to which experimental animals were exposed and on the numerous and often irreconcilable assumptions underlying their respective risk assessments.

In cancer (animal) bioassays, toxicologists try to limit the highest dose to an MTD that causes no more than 10% weight loss in the animals, limited toxic effects, and few early deaths. In the earliest (1973) Toth studies, animals were fed high doses of UDMH (29 mg/kg/day). Many animals died early and the study would not be considered valid for risk estimation today.[8] Performing cancer bioassays at such high levels can be misleading because tumors may arise from toxicity mechanisms rather than from "any inherent tumor-causing potential of the test chemical."[5] High doses of a chemical can kill cells, which results in the rapid production of new cells. The time of rapid cell growth is when the body is at greatest risk for "cancer-initiating events."

Beginning in January 1986 Uniroyal was required by the EPA to conduct toxicity and carcinogenicity studies in order to continue use of Alar. The Uniroyal studies showed no excess cancer incidence from Alar in mice or rats. Mice showed no tumors from UDMH when they received it at the maximum levels below toxicity, which were 2.9 mg/kg/day for males and 5.8 mg/kg/day for females. This was 35,000 times higher than the highest estimated daily intake for preschoolers. When mice were given amounts above the toxicity threshold they did exhibit excess tumors. This level was 23 mg/kg/day (11 blood vessel and lung tumors). At this level 80% of mice died prematurely from toxicity.[5]

EPA's new risk assessment figures announced in early 1989 were derived from data from this study. It had been assumed that mice receiving 11.5 mg/kg/day would have a statistically significant (increased) tumor incidence by the end of the study. In fact, they did not. From the numbers in the 23 mg/kg/day group, the EPA concluded that the use of Alar would result in the lifetime risk of 45 cancers per million exposed persons. The EPA typically bans any agricultural chemical that results in more than one cancer per million exposed persons.

With the announcement that mice exposed through drinking water to 23 mg/kg/day of UDMH were dying early from tumors, the EPA offered qualifying information that these deaths may be due to excessive toxicity. However, the public did not receive this information, nor did it realize that the EPA's ban was based on one study involving doses of UDMH that were almost as high as the amounts in the Toth studies that had been discarded.

In other studies conducted by Uniroyal in the mid-1980s, rats and mice were given up to 20 ppm of UDMH in water. These produced no increased incidence of tumors.[9] The EPA then required that the levels be raised to 40 and 80 ppm (7 and 13 mg/kg/day, respectively). Uniroyal contended that such high doses were above the MTD and could not be used to predict cancer risk. Although the studies showed excess lung and liver neoplasms seemingly related to UDMH administration, at these carcinogenic doses there were also deaths, hepatic toxicity, and toxicity to the hematopoietic system.[8,10] According to Uniroyal (and commonly agreed upon by the scientific community), this clearly indicated toxic levels.

The NRDC had its own set of assumptions quite different from those of EPA or Uniroyal. Its potency estimate of 8.9 was taken from the 1973 Toth study that indicated cancer potency 10 times greater than that used by the EPA studies of the mid-1980s. The NRDC claims that the figure was taken from a list of potencies for pesticides published by the EPA in

1987. But there is some confusion here because the EPA declared the Toth studies to be inadequate.[5,6]

The NRDC also used a time-dependent mathematical model that assumes that exposure to a genotoxin is more serious earlier in life because the cells have more time to multiply, thus children are more sensitive than adults. This also assumes that Alar and UDMH are genotoxins, an assumption that has not been established. The EPA, by contrast, used a time-independent model in which age at the time of exposure made no difference.

The EPA and NRDC also used different estimations of consumption. The EPA used data from a 1977-78 survey of 30,000 people performed by the U.S. Department of Agriculture (USDA). The NRDC used a later survey from 1985-86 of 2,000 people, which estimated a 30% greater level of consumption.

A study by the California Department of Food and Agriculture (CDFA) used a combination of EPA and NRDC figures and assumptions and came up with conclusions different from either of them. The CDFA used the 1977-78 survey along with recent cancer bioassays, similar to the EPA's, but it differed in that it factored in results of studies showing no cancer risk. The CDFA did use a formula that assumed a greater risk for children, although significantly less than the NRDC's estimate. The CDFA risk estimate using these assumptions was 2.6 excess cancer cases per million persons. It was further pointed out that this risk could be reduced to less than one per million by eliminating the use of Alar on apples used for applesauce or apple juice.

Also involved in assessing Alar's risks was the United Nations WHO and Food and Agricultural Organization (FAO). The FAO/WHO joint panel on pesticide residues in food carried out its own daminozide and UDMH studies on mice, rats, rabbits, guinea pigs, and dogs. The group concluded that the acceptable daily intake for humans was up to 0.5 mg/kg/day for daminozide containing less than 30 ppm UDMH. This peer-reviewed recommendation represented a higher tolerance than any of the other agencies or organizations described above.

The European scientific community has ignored the Alar hysteria. England's Ministry of Agriculture, Fisheries and Food reviewed the daminozide and UDMH toxicology and exposure data bases and in December 1989 issued a statement that "there is no risk to consumers from the use of daminozide." The ministry concluded that even for infants and children who consume large quantities of apple products, there is no risk from UDMH.[9]

What is the end of this story? Interestingly, EPA's interim conclusion from studies in progress when the Alar incident got started in 1989 was that UDMH is carcinogenic with a potency factor of 0.88. This was less than the value from Toth's data, which was approximately 9. However, the final conclusion, announced in 1991, classifies daminozide and UDMH as probable human carcinogens, but reduces the potency factor by roughly half to 0.46.[8,12]

CONCLUSION

In the end, this Alar story highlights the critical and uncertain line between risky and nonrisky commercial products. It represents the ongoing conflict between interest groups to magnify or diffuse risk perceptions in their attempts to marshall public opinion for or against regulation. In this case the NRDC, an environmental activist organization, did a masterful job of molding public opinion, overwhelming the defenses of the apple growers and the Uniroyal company and prevailing in its quest for elimination of the chemical and publicity for itself. The end of this story may be played out in the courts as the apple growers attempt to recover damages from the NRDC for loss of apple sales during the intense period of this public debate. This saga illustrates the paucity of data that may spark such fervor. It raises to new heights the ways in which these data may be distorted and expressed to marshall public support. Finally, it bespeaks a crying need for public education to ensure more informed understanding of the line between science and perception and their regulatory responses.

REFERENCES

1. Toth, B., 1,1-Dimethylhydrazine (unsymmetrical) carcinogenesis in mice. Light microscopic and ultrastructural studies on neoplastic blood vessels, *J. Natl. Cancer Inst.*, 50, 181, 1973.

2. *Federal Food, Drug, and Cosmetic Act, as Amended, and Related Laws*, U.S. Government Printing Office, Washington, DC, 1990, 33.

3. Toth, B., Induction of tumors in mice with the herbicide succinic acid 2,2—dimethylhydrazide, *Cancer Res.*, 37, 3497, 1977.

4. Toth, B., The large bowel carcinogenic effects of hydrazines and related compounds occurring in nature and in the environment, *Cancer*, 40, 2427, 1977.

5. Rosen, J.D., Much ado about alar, *Issues Sci. Technol.*, 7, 85, 1990.

6. Zeise, L., Painter, P., Berteau, P.E., Fan, A.M. and Jackson, R.J., Alar in fruit: limited regulatory action in the face of uncertain risks, in *The Analysis, Communication, and Perception of Risk,* Garrick, B.J. and Gekler, W.C., Eds., Plenum Press, New York, 1991.

7. U.S. Environmental Protection Agency, *Daminozide Special Review Technical Support Document - Preliminary Determination to Cancel the Food Uses of Daminozide*, U.S. Environmental Protection Agency, Washington, DC, 1989.

8. Marshall, E., A is for apple, alar, and...alarmist? *Science*, 254, 20, 1991.

9. Uniroyal Chemical Company, Response to the Ministry of Health, Republic of Costa Rica, *Technical Report No. 11, Review of Daminozide (B-Nine)*, 1991.

10. Uniroyal Chemical Company, letter to Janet Auerbach, U.S. EPA Office of Pesticide Programs, 1990.

11. FAO/WHO, *Pesticide Residues in Food-1989, Evaluations 1989, Part II-Toxicology*, Food and Agricultural Organization of the United Nations, Rome, 1990.

12. U.S. Environmental Protection Agency, *Memo on the Third Peer Review of Daminozide and its Metabolite/Breakdown Product of 1,1-Dimethylhydrazide*, 1991.

17
THE BALANCE BETWEEN RISK AND REASON

When Perrier, the natural sparkling water, was found in 1990 to contain 15 ppb of benzene, three times the allowable drinking water limit of 5 ppb, the company promptly removed the product from the shelves. That satisfied an immediate concern, but left unfinished business for the Perrier set. "What risks are in store for me from this contaminated drink?" they wondered. After all, if 5 ppb is the upper limit, then it stands to reason that 15 ppb must be dangerous.

The public views the EPA standards as sacrosanct. Below the number is safety, above it is danger. The agency is hard pressed to dispel this inaccurate notion, even when a worried group of exposed citizens desperately needs reassurance. How, after all, can the EPA reassure the public that 15 ppb of benzene is all right to drink without undermining public confidence in their own standard of 5 ppb and the process by which standards are set? A public health body, the EPA was never designed to confront concerns about personal risks, the kind about which people actually care.

The notion that 5 ppb is the "safe" level—the level set by the regulatory agencies—but that 15 ppb is not harmful to an individual may seem incompatible, yet it is likely true. This seeming paradox highlights the line between the science of toxicology and the regulation of toxic substances.

In the summer of 1991 I met with a group of parents who were concerned about the fact that their children were exposed to petroleum-based solvents following a contamination of their school building. After I had explained the elements of toxicology, the nature of these chemicals, the levels found, and why they would not produce health problems, one parent, breathing a sigh of relief, asked, "Where were you 10 months ago when this all started?" During those 10 months they had met with engineers and NIOSH investigators, all of whom shared the numbers, but none of whom was able to put the risks into a context that was meaningful to these concerned parents.

In a subsequent activity at the same school district, some of that solvent had been spilled near and under a school building. The state Department of Environmental Resources wanted an extensive remediation, dozens of truckloads of dirt carted away and a large portion of the school torn down: total price tag $800,000. Simpler, far less expensive, yet equally health protective methods were available and practical, but getting the department to accept those alternatives required popular support. The department, after all, is a political entity. In order for the community to participate in this decision, they had to be presented the costs, the risks, and the benefits of the various alternatives: very much like the informed consent a doctor gives before recommending a major operation. The community knew that the cost of total cleanup would be a direct out-of-pocket expense for the taxpayers. Once they were able to weigh the alternative methods and to consider the trivial incremental risks entailed, they readily opted for a less-costly approach.

Risk communication is a growing problem in this new age of chemical awareness. Rather than be assisted, it is actually impeded by this difficult-to-comprehend line between public policy and medical science. People learning that their water or air has been contaminated with a possible cancer causer are understandably frightened. Lest we suffer societal anxiety over these matters, the public must be properly informed. If the public has a right to know, surely it has the right to know the difference between known and theoretical hazards. Without such an understanding, we can never explain to the Perrier drinker what the 15 ppb means to him or her or help the citizens of a concerned community to decide how deeply into their pockets they must reach to preserve their health.

The professional discipline of risk communication is a growing one. Authors such as Covello, Sandman, Wilson, Harrison, and others have contributed in their professional activities and writings to an understanding of the ways in which people learn, accept, and process information about environmental risks.[1-7] They and others of us interested in this field face a significant future challenge: framing a clearer public understanding to guide rational policy.

Many imperatives mandate a public understanding of these distinctions between potential and known risks. Budgetary limitations are demanding a closer look at the costs of environmental cleanup. Asbestos in schools, landfills, harbors, and rivers cannot be cleaned to the point of zero contamination, but only to some point. Public participation in fiscal and health decisions requires a clear understanding of regulatory policy, health risks, and potential versus proven hazards.

At this writing, of nearly 1,245 sites on the National Priority List of Superfund sites, only 34 have been cleaned up. Approximately 40% of the money invested in CERCLA has gone to administrative oversight and litigation costs. The protracted cleanup, massive costs, and enormous transaction expenses are related to both the broad assignment of responsibility and to unfocused and unreasonable excesses of mandated cleanup. The EPA set unattainable requirements "demanding that, in many cases, a toxic waste dump should have its soil sufficiently clean so that a well producing potable water could be dug in the middle of it, regardless of whether the land was to be under a factory or out in the desert, where it posed no threat to a surrounding population."[8] It has been estimated that if real health threats and risks actually associated with these sites had been determined and had the cleanup been matched to those risks, that 90% of the sites could have been cleaned up by now for the money spent on 3%. It is, and will be more so in the future, a fiscal impossibility to return land and water to prehuman status. Priorities will have to be based upon less idealistic and more realistic goals. Those goals will of necessity be related to potential health threats to populations at risk, threats that will necessarily be assessed through better science and less emotion than they have in the past.

Thus, setting such priorities is not merely a desirable goal, it is absolutely essential. Unfocused cleanup efforts are expensive and are diverting resources from the broadest good to a far more narrow impact. Similarly, Wildofsky and many other argue that unrestrained environmental regulation has a disastrous impact upon our growth, prosperity, competitiveness, and, yes, ultimately upon the public health worldwide.[9]

In *The Structure of Scientific Revolutions,* Thomas Kuhn introduced the concepts of paradigms and their influences upon scientific thought.[10] Kuhn recognized that each discipline has its paradigms, basic hard-wired concepts that comprise the essential belief systems of those disciplines. This book has been about distinguishing the paradigms of scientific toxicology from those of regulatory toxicology. More important, it has been about letting other disciplines and the public as well in on the story.

The basic science of toxicology, as all of the basic sciences, operates under the paradigm that coherent natural laws guide the function of the universe, that these laws can be examined through experimentation, and, moreover, that absent such experimental evidence, those laws cannot be understood.

Rooted more in the fabric of public policy than in the substrate of science, regulatory toxicology operates under an entirely different set of paradigms. The most influential of these is that risks are to be avoided at all costs and that all risks are avoidable if enough money is spent. Both of these are, I believe, destructive. In *Technological Risks,* H.W. Lewis notes, "We are obsessed with risk, especially in the novel forms brought to us by science and technology. Risk has become a major political and social issue, provoking widespread uneasiness about scientific progress; demagogues thrive in such an atmosphere."[11] Fairlie called "America's morbid aversion to risk," the fear of living.[12] The fear of environmental risks is the central paradigm underlying public perceptions and the regulatory response to those perceptions. Subparadigms include many that we have discussed in preceding chapters: if one can measure it, it must be eliminated; if animals respond to high doses, humans will respond to low doses; there are no thresholds for carcinogens; if safety cannot be proven, risk must be assumed; man-made is more dangerous than natural; our health is worse or more precarious than it used to be. Some of these paradigms are held as political accommodations enabling action in the absence of knowledge; others are purely the product of Lewis' demagogues who use uncertainties to intensify fears.

For those pressing for intensified environmental controls, the receding zero has supplemented the concept of guilty-if-innocence-cannot-be-proven as a basic paradigm. Together, these perceptions argue for perpetual risk and an ongoing need to remove every last identifiable molecule. The problem with that notion is that it makes little sense scientifically, it has no end, and it becomes increasingly expensive as remediation efforts target smaller and smaller quantities. Thus, we could labor vainly until eternity to eliminate the last vestige of asbestos. When the AHERA required all asbestos to be removed from school buildings, it launched an exorbitant exercise in marginal, if any, risk reduction. Science does not tell us that ambient school building asbestos levels pose a threat. However, the guilty-unless-innocence-is-proven concept, combined with the receding zero, produce fear, perceived danger, and political pressure to respond.

For the past 20 years, these chemical risk paradigms, accepted as nearly immutable truths, have formed our federal cancer policy and have contributed to both perceptions about chemicals and cancer and to regulation of those chemicals. None of these—that animal carcinogenicity is a clear indicator of human carcinogenicity, that dose is irrelevant, or that there is no threshold for a carcinogen—is a scientific truth. Furthermore, as the saccharine story illustrates, these axioms are not even immutable perceived truths. When they oppose the public interest, the perception of their validity and relevance quickly changes. In the case of saccharin, people wanted it and therefore ignored any unknown risks.

CONCLUSIONS

In the next 10 years, attitudes will inevitably change as people are asked to pay for impossible cleanup. Low-level asbestos in public buildings and minute levels of various waste chemicals raise immense fiscal considerations. But if the public is to accept the inevitable fact that elimination of all chemicals and eradication of all risk is not possible, they must be let in on the regulatory and the scientific paradigms. They must be helped to understand the differences between perceptions, regulatory assumptions and scientific truths. It is ultimately the responsibility of all of us with professional involvement in these fields—scientists, physicians, industrial hygienists, regulators, legislators, and attorneys—to help facilitate these paradigm shifts. People cannot be left to believe erroneously that environmental cleanup comes down to two either-or choices: bankruptcy or universal cancer. Rather, there is wide room for maneuvering and for making decisions that are effective from both financial and health standpoints. If the public sees only two possibilities: trading health for money, we are in for societal trouble. The inevitable result will be confusion, widespread public anxiety, the rise of medical quackery, vastly misdirected energies, and a loss in confidence in our public institutions.

REFERENCES

1. Covello, V.T., Communicating right-to-know information on chemical risks, *Environ. Sci. Technol.*, 23, 12, 1989.

2. Sandman, P., *Explaining Environmental Risk,* U.S. Environmental Protection Agency, Office of Toxic Substances, U.S. Government Printing Office, Washington, DC, 1984.

3. Covello, V.T., Sandman, P. and Slovic, P., *Risk Communication, Risk Statistics, and Risk Comparisons,* Chemical Manufacturers Association, Washington, DC, 1988.

4. Wilson, R. and Crouch, E.A., Risk assessment and comparisons: an introduction, *Science,* 236, 267-270, 1987.

5. Covello, V.T., von Winterfeldt, D. and Slovic, P., Communicating scientific information about health and environmental risks: problems and opportunities from a social and behavioral perspective,

in *Risk Communication,* Davies, J.C., Covello, V.T. and Allen, F.W., Eds., The Conservation Foundation, Washington, DC, 1986.

6. Hance, B.J., Chess, C. and Sandman, P.M., *Industry Risk Communication*, Lewis Publishers, Boca Raton, FL, 1990.

7. American Chemical Society, *Chemical Risk Communication: Preparing for Community Interest in Chemical Release Data,* American Chemical Society, Washington, DC, 1988.

8. Koshland, D.E., Toxic chemicals and toxic laws, *Science*, 253, 949, 1991.

9. Wildovsky, A., Playing it safe is dangerous, *Regul. Toxicol. Pharmacol.*, 8, 283, 1988.

10. Kuhn, T.S., *The Structure of Scientific Revolutions,* University of Chicago Press, Chicago, 1970.

11. Lewis, H.W., *Technological Risks*, Norton and Co., New York, 1990.

12. Fairlie, H., Fear of living, *New Republic*, 14, 1989.

INDEX

Absorption. *See* Chemical absorption
Acute toxicity
 defined, 54
 dioxins, 176
Agency for Toxic Substances and Disease Registry, 224
Agent orange
 litigation, 154-155, 157
 media focus on, 171-172
 Vietnam veterans and, 171-172, 178
Agricultural chemicals. *See* Pesticides; *specific chemicals by name*
AHERA. *See* Asbestos Hazard Emergency Response Act of 1986
Air purity, 108. *See also* "Sick building syndrome"
Alar
 perceptions of risk, 247-248, 250-251, 253-254
 public relations campaign against, 247-248, 250-251, 253-254
 regulatory history, 248-250, 251
 risk assessments, 150-151, 247, 250, 251-254
Aldrin, 6, 134
Allergy and allergic responses, 47, 53
Ames, Bruce, 97, 102, 111, 115, 118
Ammonia, 54
Analytical epidemiological studies, 65-66
 case-control studies, 68, 69
 cohort studies, 68, 70-72
 considerations, 71
 interventional studies, 68
Anecdotal observations
 experimentation, 17-18
 symptoms developing from, 35
Animal studies and testing
 dioxin exposure, 175-176
 relevance to humans, 93-94, 146-148, 175-176
 use of data in risk assessment, 149-151
Anti-scientific sentiment, illness and, 21-23
Asbestos
 fiber type, 210, 210-211, 211-215
 Health Effects Institute Asbestos Research report, 213-214
 litigation, 155-156, 216-219

Asbestos (continued)
low levels of, 211-212
lung disease and, 52, 209-215, 217
occupational exposure, 210-211, 155-156
OSHA exposure level standard, 210-211
public policy debate, 215-217
"revisionists" vs. "third wavers," 215-217
risk assessment, 211, 211-215
Asbestos Hazard Emergency Response Act of 1986, 210, 216-217
Asbestosis, 54-55, 209-210, 210-211, 214-219
ATSDR. *See* Agency for Toxic Substances and Disease Registry
Attorneys. *See also* Lawsuits
mass torts and, 159
Autointoxication disease theory, 16

Bacon, Sir Francis, 15
Basic science, 28. *See also* Science
"Basic science forcing," 95, 132
Behavior. *See* Fear of contamination
Benzene, 52, 259
Bias
influence of, 32, 33
minimizing and controlling, 31-32
Biological truths, 15-24
Blinding, 32
double blinding, 32-35
Bloodletting, disease theory and, 17
Bolander, K., 101, 102-103, 103-104
Brain damage, lead poisoning, 229-231
Brain injury, toluene, 56-58
Broad Scan Study, 110-111

California Department of Food and Agriculture, 252-253
Cancer, 58-61
asbestos exposure and, 210-215
chemical risk paradigms, 259-260
"environmental origin," 91-92, 104-105
occupational, 79-80
regulatory policies, 132-137
Cancer causation, mathematical modeling and, 144-145
Cancer rates, 116-117
Candidiasis. *See* Chronic candidiasis syndrome
Carbon monoxide, 52-54, 198-199, 199-200

Carcinogenesis, principles of carcinogenic hazard, 133-135
Carcinogens, 6, 58-61, 104-104. *See also specific chemicals*
  assessment guidelines, 133-137
  HERPS index, 116
  man-made vs. natural, 111-112, 115
  the media and, 96-98
  research, 93-94
Carson, Rachel, 87-88
Case-control studies, advantages and disadvantages, 71-72
Case reports/case series, 67
CDC. *See* Centers for Disease Control
Centers for Disease Control
  identifying Legionnaires disease, 190-191
  lead exposure standard, 230, 232, 238
Chemical absorption, 49-50
Chemical bans, DDT, 135-137
Chemical hysteria. *See* Fear of contamination
Chemicals. *See also* Industrial chemicals and solvents; *specific chemicals by name*
  adverse effects, 52-54
  cancer and, 94-95, 116-117
  HERPS index, 118
  individual susceptibility to, 53-55
  measurement by parts per billion, 109-112
  misinformation, 97-98
  physical processing of, 50-51
  "the receding zero" concept of purity, 108-110
  risk paradigms, 260
  sick building syndrome and, 187-188, 190-191, 199-202
Chemicals, cancer causing. *See* Carcinogens
Chlordane, 6
Chloroform, 248
Chronic candidiasis syndrome, 21-22, 35-36
Chronic toxicity
  defined, 54-55
  limitations of data, 55-61
Claus, G., 101, 102, 103
Clean Air Act, 121, 133
Clean Water Act, 121
Clinical ecologists. *See* Environmental physicians
Clusters of disease and illness, 159-160
Cohort studies, 68-69, 70-72
  advantages and disadvantages, 71-72

Combustion products, 198-199
Common beliefs. *See* Scientific knowledge vs. common beliefs
Commoner, Barry, 88
Comparative toxicity, 40, 42
Compensation for exposure. *See also* Lawsuits
  asbestos-related disease, 55
Comprehensive Environmental Response, Compensation and Liability Act, 121
Confounders, minimizing in experimentation, 31-32, 33, 34
Consumer Product Safety Commission, 190-191
Contamination. *See also* Chemical absorption; Exposure
  health concerns, 2-4
  perception of, 1-8, 154-155, 157
  and "the receding zero" concept of purity, 108-110
Controlled studies, experimentation, 15-18
Controls, 20-21, 29, 31
Cornforth, Maurice, 100
Correlation studies, 66-67
CPSC. *See* Consumer Product Safety Commission
Cross-sectional surveys, 67-68
Cultural factors, 32
Cyanide, 52

Daminozide. *See* Alar
Data analysis, and experimental design, 36, 37
Davis, Devra Lee, 95, 132
DDT, 6, 132-136
Delaney clause of Food, Drug, and Cosmetic Act, 248-249
Descriptive epidemiological studies, 65
  case reports/case series, 67
  cross-sectional surveys, 67-68
  population studies, 66-67
Dieldrin, 134
Dietary carcinogens, 111-112, 115. *See also* Foods
1,1-dimethylhydrazine. *See* Unsymmetrical dimethyl hydrazine
Dioxins, 6-7, 148. *See also* Agent orange
  carcinogenic propensity, 115
  human epidemiological studies, 176-181
  perception of, 172
  relative toxicity, 175-176
  safety margins, 181-182
  sources of, 173-175
  structure of, 173

Disease causation
analysis principles, 163-164
causal relationship assessment, 72-74, 162-166
and the environment, 22-23
fear of chemicals and, 156-158
Dosage
poisoning and, 40, 41
toxicity and, 39-47
Dose-response relationship, 42-47, 61, 109-110, 163-164
Double blinding in experiments, 32-35
Drinking water, 50. *See also* Water purity
lead in, 224, 225, 2228-229
Drug regulation, 140-141
Dubos, René, 88
Dust, lead containing, 224, 225-228

"Ecological fallacy" concept, 66
Efron, Edith, 103
Electric and magnetic fields, 78-79, 81
Empirical observations. *See* Anecdotal observations
Environment
disease causation, 22-23
"environmental" origin of cancer, 91-92, 116-117
environmental sources of toxins, 50
Environmental concerns, 87-89
Environmental hazards
labeling as, 6
risk focus, 121-122
statutory requirements and categorization, 122, 132
Environmental illness, 22-23, 53-54
Environmental physicians, 3-4, 22, 53-54, 156-157
Environmental Protection Agency, 89, 121
Alar regulation, 248-249, 251, 252, 253
Broad Scan Study, 110-111
emergency response, 2-3, 6
public perception of standards, 257
Environmental scientists, 98-105
EPA. *See* Environmental Protection Agency
Epidemiological studies, 19
analytical, 65, 68-72
causal relationship assessment, 72-74, 162-164
descriptive, 65, 66-68
experimental inaccuracies, 81

Epidemiological studies (continued)
human epidemiology of dioxins, 176-181
quality and reliability of, 81
toxic tort testing, 165
use and misuse, 74-80
Epidemiology
definition and purpose, 65
and disease causation, 72-74, 162-164
Epstein, Samuel, 104
Erlich, Paul, 88
Erroneous beliefs and practices, 20-21
Errors in quantitative risk assessment, 149-150
Ethyl alcohol, 40
Experimental design, 18-19, 164
comparison in, 31
considerations, 37
data analysis and, 36
minimizing bias, 31-32
in waste site epidemiological studies, 74-77
Experimentation
anecdotal observations, 17-18
blinding, 32
controlled studies, 15-17
controls, 29, 31
direct experimentation, 18, 27-29
double blinding, 31-32
independent and dependent variables, 29, 30
influence of biases, 32, 33
minimizing bias and confounders, 31-32, 33, 34
observational studies, 19, 27-29
statistical analysis, 35-37
Expert testimony and witnesses, 160-166
Exposure. *See also specific toxins and chemicals by name*
clinical data, 7-8
context of, 111, 115
and health complaints, 4-5, 22-23
protection limitations, 44, 45
risk assessment, 143-151

Fairlie, H., 260
FAO. *See* Food and Agricultural Organization
Fat, PCBs in, 3, 4-6, 7-8, 9

Fear of contamination, 1-8
  cause of illness and claims, 156-158
"Fear of Living," 264
Federal Insecticide, Fungicide and Rodenticide Act, 121, 249
Fenton Communications, 250, 255
FIFRA. *See* Federal Insecticide, Fungicide and Rodenticide Act
Fiscal issues, 258, 261
Food, Drug and Cosmetic Act, 31, 121, 248-249
Food and Agricultural Organization, 253
Foods
  Delaney clause of Food, Drug and Cosmetic Act, 248-249
  HERPS index of carcinogenicity, 118-119
  lead in, 224, 225, 227
  naturally occurring carcinogens in, 111-112, 115
  naturally occurring chemicals in, 97-98
Formaldehyde, 187, 191, 199
Fringe medical groups, 3-4

Gasoline, 224, 225
Genetic defects, agent orange and, 171-172
Genotoxic process, 59-61
Glue sniffing, 56-57

Health concerns
  contamination, 2-4
  and rise in regulation, 121-141
Health Effects Institute Asbestos Research report, 212-215
"Healthy worker effects," 201
HEI-AR. *See* Health Effects Institute Asbestos Research report
Heisenberg, Werner, 100
Higginson, John, 91-92
Hill's causation criteria, 72-73, 162, 164
Historical precedents, medical, 15-24
Human epidemiological studies, dioxins, 116-181
Humors, disease theory and, 17

Illness, anti-scientific sentiment and, 21-23
Immunological tests, 163
Index risk levels, regulatory agencies and, 146
Individual susceptibility, 53-54
Indoor air quality. *See* Sick building syndrome

Industrial chemicals and solvents
individual susceptibility to, 53-54
regulation of, 139-140
toxicity, 40-47
Ingestion and inhalation of chemicals, 49-50
Interventional studies, 68
IQ and lead poisoning, 231-233, 237-238

Job satisfaction and symptomatology, 192, 196-197
Johns Manville, 217

Keene Corporation, 217
Koch, Robert, 18
Koch's Postulates, 18
Koelle, George, 108
Kuhn, Thomas, 259

Laboratory testing, toxic tort testing, 165
Landrigan, Philip, 75-76
Lane, Sir William Arbuthnot, 16-17
Laws. *See* Regulation
Laws of averages, 36-37
Lawsuits
asbestos litigation, 216-218
burden of proof, 161
chemical fears as cause of illness and claims, 156-158
and clusters of disease and illness, 159-160
economic strife and, 159
expert testimony and witnesses, 160-166
"mass torts," 153, 159
public outrage as cause, 158-159
questionable claims, 155-156
rise in, 153-154, 155
science-for-hire, 161-162
toxic tort testing, 165
Lead, 52
Lead dust, 234, 235-237
Lead exposure and poisoning, 55
CDC standard for exposure, 223, 228-229, 232, 237
chronic toxicity, 55-56
diagnosis of, 56
difficulties in determining exposure, 233-237
effects in children, 229-232

levels of significance, 228-229
limitations of data, 223-224, 233-237
low levels and intellectual function, 231-233
perceptions of, 223, 237
sources, 224-229
Lead paint, 224, 225
Legionella pneumophila bacteria, 191
Legionnaires disease, 191, 192, 193
Lewis, H.W., 260
Lind, John, 16
Litigation. *See* Lawsuits; *specific toxins and chemicals by name*
LOAELS, 44, 47
Love Canal, 159-160
Lowest observed adverse effect level. *See* LOAELS
Lysenko, Trofim, 101-102
Lysenkoism, 101-102, 103

Magnetic fields. *See* Electric and magnetic fields
Malignancy, 59, 60
Mass torts, 153, 159
Mathematical risk assessment, 143-151
Measurement of toxic levels, 109-111
Media
agent orange and, 97, 171-172
Alar and, 249-251
building-related health issues, 189-190, 191, 192
popularization of "syndromes," 189
role in war against cancer and carcinogens, 96-98
Medications
therapeutic/toxicity ratio, 39-40
toxic specificity and, 52-53
Mesotheliomas, 210, 211, 212, 213, 216
Metabolic processing of chemicals, 50-51
Methyl isocynate, 54
Misinformation
agent orange and, 172
examples, 1, 97-98
Mitlin, M., 100
Monson, Richard, 176
Mothers and Others for Pesticide Limits, 251
Multiple chemical sensitivities. *See* Environmental illness
Mutagens, 59, 61

National Cancer Act, 121
National Institute of Occupational Safety and Health, 121, 195, 199, 203, 257. *See also* Occupational Safety and Health Administration
National Resources Defense Fund, 247-248, 250-251, 253, 254
National Toxicology Program, 121
Nerve gases, 54
NIOSH. *See* National Institute of Occupational Safety and Health
Nitrogen dioxide (NO_2), 198-199
No observed adverse effect level. *See* NOAELS
NOAELS, 44, 47
NRDC. *See* National Resources Defense Fund

Observational biases, 20-21, 32
Observational studies. *See* Epidemiological studies
Occupational cancers, 79-80, 81
Occupational exposure, 58, 79-80
  asbestos, 210-211, 211-217
Occupational regulation, 137-138
Occupational Safety and Health Administration, 89, 121, 200, 211. *See also* National Institute of Occupational Safety and Health
Odors and sick building syndrome, 188-189, 196, 202-204
Organic solvents, chronic toxicity, 56-57
Organophosphates, 52
Ottoboni, Alice, 145

Pain relief, 32
Paradigms and scientific thought, 259-261
Parts per billion, measurement of toxic levels, 109-110, 109-111
Passive tobacco smoke, 198, 228
Patient bias. *See* Observational biases
PCBs. *See* Polychlorinated biphenyls
Perrier, 257
Pesticides. *See also* Alar
  Delaney clause of Food, Drug and Cosmetic Act, 248-249
  emotionalism and, 247
  EPA assessment of DDT, 133-136
  lawsuits, 156-158
  natural, 111, 115
4-phenyl cyclohexane, 188

Physical processing of chemicals, 50-51
  effects on certain organs, 52-53
Physicians
  emergency response, 3
  expert testimony and witnesses, 160-166
Placebo, 31
Poisoning. *See* Contamination; Exposure
Politically motivated science, 98-105
Pollution, 88
Polychlorinated biphenyls
  in blood, 4, 5-6, 7-8
  fat levels, 9
  illustration of contamination, 1-8
  scientific literature, 4-6
Population studies, 66-67
Proposition 65, 247
Prospective studies. *See* Cohort studies
Public opinion
  and environmental concerns, 87-89, 257-261
  fear of risk paradigm, 260
  generalizations and, 104
  influence of professionals, 261
  influence of the media on, 96-98
  lawsuits resulting from public outrage, 158-159
Public policy
  balancing risk and reason, 257-261
  fiscal issues, 258, 261
  and limitations of science, 94-95
  and risk assessment, 150-151
  setting priorities, 259
Purity, 108-110

Questionnaires, unreliability of data from, 75-76, 201-202

Randomization in experiments, 31-32
"The receding zero" concept of purity, 108-110
Regulation
  drug approval studies, 31
  purpose, 121
  risks vs. benefits, 138-141
  and science, 95, 103
Regulation of toxins
  chronic toxicity issue, 58

Regulation of toxins (continued)
regulatory response and its history, 6-7
science vs. regulation, 257, 259-261
Regulatory agencies
cancer policies and, 132-137
creation of, 121-122
establishment of, 122, 132
occupational regulation, 137-138
quantitative risk assessment and, 143-151
Regulatory scientists, 99, 103, 104-105
Regulatory toxicology vs. scientific toxicology, 136-137, 257, 259-261
Resource Conservation and Recovery Act, 121
Retrospective studies. *See* Case-control studies; Cohort studies
Risk assessment, 143-151
asbestos, 209, 211-215
examples, 145-146, 148
in tort claims, 161
Risk communication, 189, 258
Ruckelshaus, William, 132-133, 135-136
Russia, politically motivated science, 100-102

Saccharin, 260
Saffiotti, Umberto, 104, 133-135
Science
generalizations and, 104
impossibility of proving safety, 95-96
mathematical modeling and, 145-146
and the media, 96-98
paradigms and scientific thought, 259-261
politically motivated, 98-105
public policy and, 94-95, 135
and regulation, 103
role in toxicological litigation, 154-156
scientific method, 18, 27-37
"trans-science," 146-147
uncertainties, 146-147, 149-150
Scientific knowledge vs. common beliefs
cancer rates, 116-117
health risks, 118
measurement of exposure, 110-111
purity, 108-110

Scientific toxicology vs. regulatory toxicology, 136-137, 257, 259-261
Scientists, expert testimony and witnesses, 160-166
Setting priorities, 259. *See also* Public policy
Shindell, Sidney, 108
Sick building syndrome, 78, 81. *See also* Environmental illness
  building-associated symptoms, 194-195
  building-related diseases, 192-193
  case studies, 187-188, 202-204
  categories of illnesses associated with, 192-195
  causes, 195
  chemicals and, 187-188, 191, 199-201
  combustion products, 198-199
  diversity of professional opinion and response, 202-204
  investigating complaints, 201-202
  job satisfaction and, 192, 196
  mass hysteria, 197-198
  media and, 189-190, 191, 192
  odors and, 188-189, 196, 202-204
  perception of, 188-192
  psychological factors, 192, 195, 196, 197-198
  risk communication, 189
  "tight building syndrome," 193-194
  ventilation and, 192-193, 195-196
Skin absorption of chemicals, 49-50
Smoking
  and indoor air quality, 194, 198
  lead exposure, 227
Snow, John, 19
Soviet Union, politically motivated science, 100-102
Spirapolis, John, 97-98
Statistical analysis in experiments, 35-37
Streep, Meryl, 251
Studies. *See specific studies, types, or study groups by name*
Sulfur dioxide (SO_2), 198
Superfund sites, 258-259
"Syndromes," popularization of, 189. *See also* Environmental illness

TCDD. *See* Dioxins
Testing. *See* Animal studies and testing

Therapeutic/toxicity ratio of medications, 39-40
Thresholds, 146-148, 150
"Tight building syndrome," 193-194. *See also* "Sick building syndrome"
Toluene, 56-57
Tort claims. *See* Lawsuits
Toxic specificity, 52-53
Toxic Substance Control Act, 121, 122, 132
Toxic tort testing, 165
"Toxic torts," 153
Toxicity
  acute vs. chronic, 40
  adverse effects of chemicals, 51-52
  carcinogens and mutagens, 58-59
  chemical absorption, 49-50
  comparative, 40, 42
  defined, 39, 47-49
  and dosage, 39-47
  dose-response relationship, 42-47, 61, 109-110, 163-164
  effects of chemicals on certain organs, 51-52
  ethyl alcohol, 40
  individual susceptibility, 53-54
  industrial solvents, 40-47
  introduction, 1-8
  limitations of data, 55-61
  medication, 39-40
  physical processing of chemicals, 50-51
  serious vs. benign effects, 47-49
  term in context, 51
  types, 54-55
Toxicity terminology, 47-49
Toxicology, regulatory vs. scientific, 136-137, 257, 259-261
"Trans-science," 146-147
Trichloroethylene, 40, 42, 44, 45, 46, 148, 158-159

UDMH. *See* Unsymmetrical dimethyl hydrazine
UFFI. *See* Urea-formaldehyde foam insulation
Unsymmetrical dimethyl hydrazine, 247, 248, 249
Urea-formaldehyde foam insulation, 191, 199

Ventilation and sick building syndrome, 192-193, 195-196
Vietnam veterans, agent orange exposure, 171-172, 178
Vineberg procedure, 20-21

Visceroptosis disease theory, 16
VOCs. *See* Volatile organic compounds
Volatile organic compounds, and sick building syndrome, 199-201

Waste site epidemiology, 74-77, 81
Water purity, 108. *See also* Drinking water
Weinberg, Alvin, 146
World Health Organization, 253
Committee on Indoor Air Quality: Organic pollutants, 200